Additive and Multiplicative Number Theory

Additive and Multiplicative Number Theory

Cai Tianxin

Zhejiang University, China

Translated by

Tyler Ross

New York, USA

W♦ World Scientific

NEW JERSEY · LONDON · SINGAPORE · BEIJING · SHANGHAI · HONG KONG · TAIPEI · CHENNAI · TOKYO

Published by

World Scientific Publishing Co. Pte. Ltd.

5 Toh Tuck Link, Singapore 596224

USA office: 27 Warren Street, Suite 401-402, Hackensack, NJ 07601

UK office: 57 Shelton Street, Covent Garden, London WC2H 9HE

Library of Congress Cataloging-in-Publication Data
Names: Cai, Tianxin, 1963–   author
   http://id.loc.gov/authorities/names/no2007125298
   http://id.loc.gov/rwo/agents/no2007125298
Title: Additive and multiplicative number theory / author Tianxin Cai, Zhejiang University, China.
Description: New Jersey : World Scientific, [2026] | Includes bibliographical references and index.
Identifiers: LCCN 2025005550 | ISBN 9789819806546 hardcover |
   ISBN 9789819806553 ebook | ISBN 9789819806560 ebook other
Subjects: LCSH: Number theory
   http://id.loc.gov/authorities/subjects/sh85093222 | Numbers, Prime
   http://id.loc.gov/authorities/subjects/sh85093218
Classification: LCC QA241 .C254 2025 | DDC 512.7--dc23/eng/20250625
LC record available at https://lccn.loc.gov/2025005550

British Library Cataloguing-in-Publication Data
A catalogue record for this book is available from the British Library.

For any available supplementary material, please visit
https://www.worldscientific.com/worldscibooks/10.1142/14140#t=suppl

Desk Editors: Soundararajan Raghuraman/Rok Ting Tan

Typeset by Stallion Press
Email: enquiries@stallionpress.com

Preface

The essence of mathematics lies in its freedom.

—Georg Cantor

"In about 1940, he (Hua Luogeng) completed the composition of *The Theory of Stacked Prime Numbers* (堆类素数论) in the course of eight months." These words appear in the fifth chapter (entitled *National Southwest Associated University*) of *Hua Luogeng (A Biography)*, by the Chinese mathematician Wang Yuan; at the time in question, Hua Luogeng (1910–1985) was teaching in the mathematics department of Southwest United University in Kunming. This book was first published however in the Soviet Union in 1947, as monograph No. 22 of the Steklov Institute of Mathematics, and the first Chinese edition did not appear until 1953, when it was published by the Chinese Academy of Sciences in Beijing. The English version was published by the American Mathematical Society in 1965, and subsequently there have also been editions in German, Hungarian, and Japanese. Except for the Chinese and Japanese editions, the title of this book in other languages has always been rendered as *Additive Number Theory*.

The theory of stacked prime numbers is a bit of terminology coined by Hua Luogeng, who ingeniously inserted prime numbers in place of the additive subdiscipline of number theory. Additive number theory in turn is an important branch of analytic number theory; its central concern is the problem of representing an integer as a sum of integers subject to some specific conditions (differences of course also belong to the additive discipline), or, in general, the problem

v

of integer representations. Some important examples are Goldbach's conjecture, the twin prime conjecture, Waring's problem, the theory of integer partitions, the problem of representations of integers as sums of squares, and so on; the famous sieve method and circle method also belong to this category. Analytic number theory, on the other hand, refers to the study of problems concerning integers using tools from calculus and complex analysis, and can be divided into two main categories: additive number theory and multiplicative number theory.

The centerpiece of multiplicative number theory is the consideration of the distribution of prime numbers via the properties of multiplicative generating functions, the most famous results in this direction being for example the prime number theorem and Dirichlet's theorem on primes in arithmetic progressions. In 1967, Harold Davenport (1907–1969), a mathematician at the University of Cambridge, published a book with the title *Multiplicative Number Theory* with the Markham Publishing Company in Chicago. Unfortunately, he died two years later; I read the second edition, in which one of his students, Hugh L. Montgomery of the University of Michigan, rewrote the last seven chapters, and published them in 1980 through the prestigious Springer Verlag publishing house.

In addition to discussing the prime number theorem and Dirichlet's theorem on arithmetic progressions, *Multiplicative Number Theory* also deals with some other prime number distribution problems, the problem of distribution of zeros of Riemann zeta and L functions (including the Riemann hypothesis and generalized Riemann hypothesis), and similar topics. The methods used include, in addition to the properties of zeta, L, and Γ functions, Gaussian sums, cyclotomic fields, Dirichlet's class number formula, the Polya-Vinogradov inequality, the large sieve method, Siegel's theorem, Bombieri's mean value theorem, and so on. Two Chinese mathematicians, Hua Luogeng and Chen Jingrun, appear in the index of names at the back of the book, where Chen's name is naturally related to the study of Goldbach's conjecture. Indeed, according to the definition of multiplicative number theory, Goldbach's conjecture also lies within its scope.

But although Goldbach's conjecture and the twin prime conjecture are the most famous, the most fundamental problem in the field of additive number theory is rather Waring's problem. This problem was first introduced in 1770 by the English mathematician Edward Waring (c. 1736–1798), whose book *Meditationes Algebraicae*, published in the same year, contained the four-squares theorem, proved by the French mathematician Lagrange, which states:

> Every positive integer can be expressed as a sum of squares of four integers.

Waring further asserts that every positive integer can be expressed as a sum of nine nonnegative cubic integers, a sum of nineteen 4th powers of nonnegative integers, and more generally, he states

> Given any positive integer k, there exists a positive integer $s = s(k)$ such that every positive integer n can be presented as a sum of s kth powers of non-negative integers, symbolically
>
> $$n = x_1^k + x_2^k + \cdots + x_s^k.$$

This is the so-called Waring problem, which stood for more than two and a half centuries. It was answered with an affirmative proof in 1909 by David Hilbert (1878–1943), a German mathematician and one of the two founders of the Göttingen school of mathematics, so that this result has also been known as the Hilbert-Waring theorem. However, the research on the Waring problem was far from complete.

In Chapter 1, one related problem that has continued to interest number theorists from different countries and different periods is: given any positive integer $k > 1$, what is the smallest $s = g(k)$ that satisfies the above condition for every positive integer n? A deeper variation on this problem asks what is the smallest $s = G(k)$, such that every sufficiently large positive integer n satisfies the above condition.

For the first problem, there has so far appeared only conjecture, suggested by numerical verification amd partial proof; for the second problem, despite the concerted efforts of many generations of mathematicians (most of whom are elites engaged in the study of analytic number theory), the results obtained have been only sporadic, neither coherent nor beautiful, and there is no hope at present of a complete solution to the problem.

Precisely because we noticed this, we began to think whether the original formulation of the problem might merit some revision or addition. There is certainly nothing wrong with a problem being profound and difficult, but it is worth considering the situation of inconsistent or unattractive results. We believe that a really good mathematical problem should be profound and abstract, but at the same time simple and beautiful. It was from this point of view that we undertook to consider and study additive-multiplicative number theory, i.e., the union of additive and multiplicative equations for integers.

The first step in this direction was in 2011, when we introduced the New Waring problem:

Let k and s be positive integers and consider the diophantine equation

$$n = x_1 + x_2 + \cdots + x_s,$$

subject to the condition

$$x_1 x_2 \cdots x_s = x^k. \tag{W}$$

We denote by $g'(k)$ (respectively $G'(k)$) that smallest positive integer s such that for any (respectively any sufficiently large) positive integer n, the conditions of the additive equation (W) are satisfied.

We found that when n is a prime number other than 2, 5 or 11 it is always possible to obtain solutions. For some examples: 3 (1, 1, 1), 7 (1, 2, 4), 13 (1, 3, 9), 17 (1, 8, 8), 19 (4, 6, 9), and so on. Continuing, we obtained a series of results and prove that $g'(k) = 2k - 1$, while giving various estimates for $G'(k)$, and also presenting the conjecture $G'(k) = k$. This conclusion is much more straightforward than the original Waring problem, but I suspect it is equally difficult.

In another direction, we present a problem posed and studied by the Polish mathematician Andrzej Schinzel (1937–2021), who proved in 1996 that the equation

$$x_1 + x_2 + x_3 = x_1 x_2 x_3 = 6$$

has infinitely many solutions in positive rational numbers. For example, (1,2,3) is an obvious solution to the above equation. This equation does not define a Waring-type problem, but also involves a combination of additive and multiplicative relations. We generalize this problem to the case of arbitrarily many variables in various forms, and prove the existence of infinitely many solutions, with examples and solution methods via the theory of elliptic curves.

Returning to the new Waring problem, although it has been almost 10 years since our above conjecture was first formulated, no one has yet been able to prove it or discover any counterexample; but it has been encouraging to see that is has also attracted the attention of colleagues such as Alan Baker, a British mathematician and Fields Prize winner, and Professor Preda Mihailescu, a Romanian mathematician and professor at the University of Göttingen, both of whom have given it high praise. In a letter to the author, Baker, an expert on the Waring problem, praised our work as "a truly original contribution". Mihailescu, in a review article on the abc conjecture in the Newsletter of the European Mathematical Society, devoted a special section to the additive and multiplicative equation, in which he dubbed it the "yin-yang equation".

Chapter 2 encouraged by seniors and peers, we applied the idea of additive and multiplicative equations to several other classical problems in number theory. In 2012, the author proposed the New Fermat problem, which asks for positive integer solutions to the system of equations

$$\begin{cases} A + B = C \\ ABC = x^n. \end{cases} \tag{F}$$

When $(A, B, C) = 1$ (and $n \geq 3$), equation (F) is equivalent to the diophantine equation of Fermat's Last Theorem:

$$x^n + y^n = z^n, \quad n \geq 3,$$

so that (F) has no positive integer solutions. It is worth mentioning that Fermat himself and later Euler proved that Fermat's Last Theorem holds for $n = 4$ and $n = 3$ in 1637 and 1753, respectively, 116 years apart.

When $(A, B, C) > 1$, on the other hand, there are sometimes solutions. For example, with $n = 4$, we find four solutions, namely (2, 2, 4), (5, 400, 405), (17, 272, 289), (47927607119, 1631432881, 49559040000), which have in common that $\gcd(A, B, C) = 2, 5, 17, 239$ (respectively) are all prime numbers. We obtain various positive and negative conclusions; for example, if $(A, B, C) = p^k$, $k \geq 1$, then (F) has no positive integer solution when $n = 4$,

$p \equiv 3(\bmod\ 8)$, and likewise when $n = 5$, $p \not\equiv 1(\bmod\ 10)$, (F) has no positive integer solutions.

In addition, we also considered the case of imaginary quadratic fields and obtained the following conclusions:

Proposition. *If* t *is a squarefree integer not equal to* $0, -1$, *such that the elliptic curve*

$$tu^2 = 1 + 4k^3$$

admits nonzero rational solutions (u, k), *then* (F) *admits infinitely many solutions* (A, B, C, x) *in the quadratic field* $\mathrm{Q}(\sqrt{t})$ *when* $n = 3$.

We make several conjectures, one of which is that

Conjecture. *If* $(A, B, C) = p$ *is prime and* n *is an odd prime, then* (F) *has no positive integer solutions. If* $(A, B, C) = p^k$, $k \geq 1$, *and* n *are odd prime numbers, then* (F) *has no positive integer solution when* $p \not\equiv 1(\bmod\ 2n)$.

It is worth mentioning that if the *abc* conjecture holds, the above conjecture is true for fixed p and sufficiently large n, where n need not be a prime number; but of course the *abc* conjecture has not yet been proved.

In Chapter 3, we turned to Euler's conjecture. In 1769, while returning from Berlin to Petersburg, Euler formulated a conjecture that attempted to generalize Fermat's Last Theorem to include greater multiplicity of terms as follows.

Euler's Conjecture. For any positive integer $s \geq 3$, the equation

$$a_1^s + a_2^s + \cdots + a_{s-1}^s = a_s^s \tag{E1}$$

has no positive integer solutions.

When $s = 3$, this is the special case of Fermat's Last Theorem, so the conjecture naturally holds. When $s > 3$, it turns out that the conjecture is false in general. But although it is a now disproved conjecture, nevertheless this problem has attracted many number theorists to work out the general case or to enumerate the counterexamples, just as people still persist in finding the prime factors or factorizations of Fermat numbers even after Fermat's prime conjecture had been disproved.

In 1988, N. Elkies [E1] of Harvard University used the theory of elliptic curves to give an infinite number of counterexamples to Euler's conjecture for $s = 4$, one of which is

$$2682440^4 + 15365639^4 + 18796760^4 = 20615673^4.$$

We will present Elkies' argument in Chapter 3, along with a work by American physicist Lee W. Jacobi and mathematician Daniel J. Madden, who in 2008 studied $s = 4$ and proved that the following Euler-type equation (with linear constraints on the terms)

$$a^4 + b^4 + c^4 + d^4 = (a + b + c + d)^4$$

has infinitely many nontrivial solutions in integers.

In 2012, the authors proposed the following system of diophantine equations inspired by the New Waring problem

$$\begin{cases} n = a_1 + a_2 + \cdots + a_{s-1} \\ a_1 a_2 \cdots a_{s-1} (a_1 + a_2 + \cdots + a_{s-1}) = b^s, \end{cases} \tag{E2}$$

where $s \geq 3, n, a_i$, b are positive integers.

This is yet another type of combined additive and multiplicative equation. It is obvious that from a solution of (E1), a solution of (E2) can be derived by simply taking

$$n = a_s^s, \ a_1 = a_1^s, \ldots, a_{s-1} = a_{s-1}^s.$$

We obtain several nice results concerning this problem via the theory of elliptic curves. In particular, we prove that for $s = 3$ and any positive integer n, the system of equations

$$\begin{cases} n = a_1 + a_2 \\ a_1 a_2 (a_1 + a_2) = b^3 \end{cases}$$

has no positive integer solutions.

It follows from this that Fermat's theorem holds in the case $n = 3$, a much more concise proof than that given by Euler. The latter made use of the method of infinite descent, the unique factorization property of the integers, and congruence relations on the ring of integers in imaginary quadratic fields.

More generally, let p be any odd prime number and consider the equations

$$\begin{cases} n = a_1 + a_2 \\ a_1^{\frac{p-1}{2}} a_2^{\frac{p-1}{2}} (a_1 + a_2) = b^p. \end{cases}$$

After some manipulation, the second equation above can be transformed into

$$u^{\frac{p-1}{z}} (u + 1) = v^p.$$

Conjecture. *There are no rational solutions to the above equation other than the trivial solutions* $(0, 0)$ *and* $(-1, 0)$.

If this conjecture can be proved, Fermat's Last Theorem can be easily recovered from it.

In Chapter 4, we discuss the problem of representation of integers as sums of integer squares. It is well known that prime numbers congruent to 1 modulo 4 are uniquely representable as a sum of squares of two nonnegative integers, which we prove in four different ways, one of which is constructive. We also present proofs of Lagrange's sum of four squares theorem and Gauss's sum of three squares theorem, which were obtained in 1770 and 1796, respectively.

Gauss's Theorem. Except for numbers of the form $4^k(8n+7)$, every positive integer can be expressed as a sum of three square integers.

Note that in particular, a number of the form $8n + 3$ cannot be expressed as a sum of three. From this, Gauss easily proves that the following Fermat polygonal number conjecture (now a theorem) holds for $n = 3$.

Fermat's Polygonal Number Theorem. When $n \geq 3$, every positive integer can be expressed as the sum of no more than n n-gonal numbers.

When $n = 4$, this is the same as Lagrange's sum of four squares theorem. Once again, the proof of the case $n = 4$ came before the case $n = 3$.

For the case $n = 3$, we introduce a proof due to N. C. Ankeny. For this purpose, we need to make use of Dirichlet's theorem on primes in arithmetic progressions and the following Minkowski's theorem on the geometry of numbers.

Lemma. *Any convex region in n-dimensional space that is symmetric about the origin and has volume greater than 2^n contains a point whose coordinates are all integers and not all zero.*

In 1752, in a letter to the German mathematician Goldbach, Euler asked:

Are there infinitely many prime numbers of the form $n^2 + 1$?

This is a very difficult question that no one has been able to answer so far.

In 2016, we found a necessary and sufficient condition for $n^2 + 1$ to be a prime number: the additive and multiplicative equation

$$\begin{cases} n = a + 2b \\ ab = \binom{c}{2} \end{cases}$$

has positive integer solutions if and only if $n^2 + 1$ is a composite number, or, equivalently, $n^2 + 1$ is a prime number if and only if the above equation has no positive integer solutions.

If similarly we consider the equations

$$\begin{cases} n = a + b \\ ab = \binom{c}{2}, \end{cases}$$

we find that there exist positive integer solutions if and only if $2n^2 + 1$ is a composite number.

In 2018, we further generalized the above equation to

$$\begin{cases} n = a + b \\ ab = tP(m, c), \end{cases} \tag{CZ}$$

where a, b, and t are all positive integers and $P(m, c) = \frac{c}{2}\{(m-2)c - (m-4)\}$ is the cth m-gonal number.

If $r_{m,t}(n)_V$ denotes the number of solutions of the equation (CZ), we consider the case $r_{m,t}(n) = 0$ corresponding to the condition that (CZ) has no solutions. In particular, consider $r_{5,1}(n) = 0$, which expresses the situation that

$$\begin{cases} n = a + b \\ ab = P(5, c) = \frac{1}{2}c(3c - 1) \end{cases}$$

has no solutions, for which we obtained the sufficient condition that 6^{n^2+1} is a prime number.

Continuing along this path we discovered a series of similar conclusions, for example

$$r_{7,1}(n) = 0 \Leftrightarrow 10n^2 + 9 \in \mathbb{P} \cup 9\mathbb{P},$$
$$r_{3,2}(n) = 0 \Leftrightarrow n^2 + 1 \in \mathbb{P},$$
$$r_{8,2}(n) = 0 \Leftrightarrow 3n^2 + 8 \in \mathbb{P} \cup 4\mathbb{P} \cup 8\mathbb{P}.$$

In general, we have the following proposition.

Proposition. If $2(m-2)n^2 + t(m-4)^2 n^2 + t(m-4)^2$ is a prime number, then (CZ) has no solutions.

Chapter 5: In 1742, shortly after his transfer from the St. Petersburg Academy of Sciences to the Berlin Academy of Sciences, Euler wrote a reply to Goldbach containing the numerical observation that any even number greater than 4 can be expressed as a sum of 2 odd prime numbers.

Goldbach had previously written to Euler to inform him of his own observation that every odd number greater than or equal to 9 can be expressed as the sum of three odd prime numbers. It is easy to see that Euler's conjecture can be directly derived from Goldbach's conjecture, and so they came to be collectively known as Goldbach's conjecture.

So far, Goldbach's conjecture has not been proved. The strongest result was obtained by the Chinese mathematician Chen Jingrun in 1966, who proved that every sufficiently large even number can be expressed as the sum of an odd prime number and another odd number with no more than two prime factors. Chen used a new weighted sieve method, a variant of the old sieve of Eratosthenes. On the other hand, as early as 1937, the Soviet mathematician Vinogradov used the circle method to prove that every sufficiently large odd number is the sum of 3 odd prime numbers.

We consider that prime numbers are factors in the multiplicative decomposition of integers; they are not naturally suited to decompositions as sums. Moreover, the distinction into cases that even and odd numbers admit presentations as sums of two prime numbers or three prime numbers seems ad hoc and inconsistent with the demands of beauty. Furthermore, as n increases, the number of presentations as a sum of prime numbers gradually increases and tends to infinity, which is rather wasteful. These can be regarded as shortcomings of

Goldbach's conjecture, and taking this into account, we settled after some exploration upon a variation giving new meaning to the familiar binomial coefficients. We define the figurate primes, numbers of the form

$$\binom{p^i}{j},$$

where p is a prime number, $p^i \geq j$ for i and j positive integers. This set contains 1, all prime numbers and their powers, and infinitely many but sparsely distributed even numbers. It can be seen that the figurate primes have the properties of both prime and figurate numbers, and their number is of the same order as the number of prime numbers in the infinite sense.

We have verified the following conjecture for numbers up to 10^7:

Conjecture. *Every positive integer greater than 1 can be presented as a \ the sum of two figurate prime numbers.*

In addition, we have obtained some other conclusions and conjectures. If we exclude 1 and the trivial case (arising from the symmetry of binomial coefficients), there is a seemingly simple but still difficult to prove conjecture:

Conjecture. *The figurate primes are all different.*

In 1900, Hilbert asked 23 mathematical problems at the International Congress of Mathematicians in Paris, of which the eighth problem is undoubtedly the most important, encompassing Goldbach's Conjecture, the twin primes conjecture, and the Riemann hypothesis. He concludes with this statement:

> "After an exhaustive discussion of Riemann's prime number formula, perhaps we may sometime be in a position to attempt the rigorous solution of Goldbach's problem,21 viz., whether every integer is expressible as the sum of two positive prime numbers; and further to attack the well-known question, whether there are an infinite number of pairs of prime numbers with the difference 2, or even the more general problem, whether the linear diophantine equation
>
> $$ax + by + c = 0$$
>
> (with given integral coefficients each prime to the others) is always solvable in prime numbers x and y."

There have not since been any specific questions or conjectures regarding the above linear diophantine equation problem mentioned by Hilbert.

After introducing the notion of figurate primes, we tried to give a clear meaning to the diophantine equation indicated in Hilbert's address, while simultaneously including Goldbach's conjecture and the twin prime conjecture. On the basis of computer search, we have the following conjecture, the latter part of which can be derived from Schinzel's hypothesis using the properties of the diophantine equation:

Conjecture. Let a and b be any given positive integers, $(a, b) = 1$; if $n > (a - 1)(b - 1) + 1$, the equation

$$ax + by = n$$

has a solution (x, y) in figurate primes; if also $n \equiv a + b \pmod 2$, then the equation

$$ax - by = n$$

admits infinitely many solutions (x, y) in figurate primes.

We next review and explore the perfect number problem, perhaps the oldest extant problem in mathematics, posing as a variation the square perfect number problem, which establishes a one-to-one correspondence with the twin prime pairs of the Fibonacci sequence from the 13th century, just as Euler proved a one-to-one correspondence between even perfect numbers and the Mersenne primes of the 17th century. This too can also be seen as a problem of additive and multiplicative number theory, where additivity is obvious and multiplicativity is expressed by the presence in the sums of divisors or divisor squares. A deeper exploration of this topic has been presented by the author in his book *Perfect Numbers and the Fibonacci Sequence*, a small portion of which is reproduced here.

In Chapters 6, we explore the *abcd* equation and the new congruent numbers. In early 2013, the author defined the following so-called *abcd* equation (its name was inspired of course by the abc conjecture).

Definition. Let n be a positive integer and a, b, c, d be positive rational numbers, the so-called *abcd* equation is

$$n = (a + b)(c + d), \tag{C1}$$

subject to

$$abcd = 1.$$

By the arithmetic-geometric inequality, we have

$$(a + b)(c + d) \geq 2\sqrt{ab} \times 2\sqrt{cd} = 4.$$

Therefore, (C1) has no solution when $n = 1, 2$, or 3. On the other hand, the

$$4 = (1 + 1)(1 + 1), 5 = (1 + 1)\left(2 + \frac{1}{2}\right).$$

The uniqueness of the solution for 4 is obvious, while the uniqueness of the solution for 5 can be proved using elliptic curves. As for $n \geq 6$, we have:

(a) When $n \geq 6$, if the *abcd* equation has positive rational number solutions, then there are infinitely many such sets of solutions; in particular, the equation $13 = x + \frac{1}{x} + y + \frac{1}{y}$ has infinitely many positive rational number solutions.

We have given several conditions and conjectures for the solution of the *abcd* equation, from which we know that there are infinitely many n such that the *abcd* equation has no solution. At the same time, we also consider the following equation

$$n = \left(a + \frac{1}{a}\right)\left(b + \frac{1}{b}\right), \tag{C2}$$

where both a and b are positive integers. Clearly, if the above equation has a solution, then the *abcd* equation must have a solution. We prove that

Proposition. *(C2) has a solution if and only if* $n = F_{2k-3}F_{2k+3}$ *($k \geq 0$), in which case the solution is* $(a, b) = (F_{2k-1}, F_{2k+1})$.

It follows that there exist infinitely many n (4, 5, 13, 68, 445, 3029, 20740, ...) such that the *abcd* equation has a solution.

But there are still infinitely many n's for which we cannot determine whether the *abcd* equation has any solution, recalling the congruent number problem. This is one of the most original problems in this book, and its research method is both subtle and rich, and immeasurably difficult. Both the determination of its solvability and the number and structure of solutions when there are solutions are problems worth studying.

Next, we explored the ancient problem of congruent numbers, which probably originated in Arabia and was studied by the mathematician and engineer Al-Karaji in Baghdad at least in the 10th–11th centuries. A congruent number is such a positive integer n that represents the area of a right triangle with all three sides having rational length. For example, 6 is a congruent number corresponding to the right triangle with sides of length (3, 4, 5).

The congruent number problem is as follows: "a simple discriminant is sought in order to decide whether a positive integer is a congruent number." Clearly, for any positive integers m and n, $m^2 n$ is a congruent number if and only if n is a congruent number. So we can restrict our inquiry to squarefree positive integers.

In the 13th century, the Italian mathematician Fibonacci proved that $n = 5$ is a congruent number corresponding to the triangle with side lengths $\left(\frac{3}{2}, \frac{20}{3}, \frac{41}{6}\right)$. Fibonacci also asserted that $n = 1$ is not a congruent number, but unfortunately his proof was erroneous. Some four centuries later, Fermat gave a correct proof. This result also leads to the fact that Fermat's Last Theorem holds for exponents of $n = 4$, a rare example of a proof given during Fermat's lifetime. It also follows that every square number is not a congruent number. In the 18th century, Euler discovered that 7 is a congruent number.

Clearly, a sufficient condition for n to be a congruent remainder is that the system of equations

$$\begin{cases} a^2 + b^2 = c^2 \\ \frac{1}{2}ab = n \end{cases}$$

admits a solution in positive rational numbers (a, b, c).

In the second half of the 20th century, mathematicians found that the problem of congruent numbers is as closely related to elliptic curves as is Fermat's Last Theorem. In fact, it corresponds to the equation of the elliptic curve given by

$$y^2 = x^3 - n^2 x.$$

Using the theory of elliptic curves, it can be shown that

for primes $p \equiv 3 \pmod 8$, p is not a congruent number, but $2p$ is a congruent number;

for primes $p \equiv 5 \pmod 8$, p is a congruent number,

for primes $p \equiv 7 \pmod 8$, both p and $2p$ are congruent.

The second result above was obtained in 1952 by Heegner, a German radio engineer, who was the first to prove that there are infinitely many congruent squarefree numbers. In 2014, Tian Ye proved that, given any positive integer k, there are infinitely many squarefree congruent numbers with exactly $k+1$ odd prime factors in each residue class of 5, 6, and 7 modulo 8. In 2017, Tian Ye, Xin Yi Yuan, and Shou Wu Zhang gave several sufficient conditions for a positive integer n congruent to either 5, 6, or 7 modulo 8 to be a congruent number, and they believe that these results can be extended to show that the classes of such integers have positive density.

On the other hand, there is Goldfeld's conjecture, which states that the congruent numbers and noncongruent numbers are equally distributed among all integers. To be precise, almost all positive integers congruent to 5, 6, or 7 modulo 8 are congruent numbers, and almost no positive integers congruent to 1, 2, or 3 modulo 8 are congruent numbers. Finally, it has been shown that assuming the truth of the BSD conjecture leads to the result that every positive integer congruent to 5, 6, or 7 modulo 8 is a congruent number. The smallest congruent number outside of this family is 34, corresponding to a triangle with side lengths $(225/30, 272/30, 353/30)$; the smallest congruent numbers congruent to 1 and 3 modulo 8 are are 41 $(40/3, 123/20, 881/60)$ and 219 $(55/4, 1752/55, 7633/220)$, respectively.

We present a variant of the congruent number problem using the idea of additive and multiplicative equations, after presenting Heegner's theorem and Tunnell's theorem. This time the variant is due to the observation that the right triangle on which the congruent number problem depends is a special case of an isosceles right-angled trapezoid. We replace the right triangle in the definition of congruent numbers with a right-angled trapezoid and consider it in three cases:

(1) congruent numbers, when both n and the upper base d are positive integers;

(2) k-congruent numbers, when the ratio k of the lower and upper bases is a positive integer;

(3) d-congruent numbers, when the upper base d is a non-negative integer.

It can be deduced that the elliptic curves corresponding to the k-congruent numbers and the d-congruent numbers are respectively

$$E_{n,k} : y^2 = x^3 - \left(k^2 - 1\right) n^2 x.$$

$$E_{n,d} : y^2 = x^3 - \frac{3n^2 + d^4}{3}x + \frac{\left(9n^2 + 2d^4\right) d^2}{27}.$$

When $k = 1$ or $d = 0$, the above elliptic curve becomes the general elliptic curve for ordinary congruent numbers.

It is not difficult to prove that almost all positive integers are generalized congruent numbers. Further, we use the analytical method and the prime number theorem to prove the following:

let $f(x)$ denote the number of non-concurrent integers that do not exceed x. Then

$$f(x) \sim \frac{cx}{\log x},$$

where $c = 1 + \ln 2$.

Using the theory of elliptic curves, we prove that

every positive integer is a k-congruent number for some k;

every positive integer is a d-congruent number for some d;

We propose the following stronger conjecture:

Conjecture. *For every positive integer n, there exist infinitely many positive integers k such that n is a k-congruent number.*

Finally, in Chapter 7 we study the additive and multiplicative congruences, we investigate a problem proposed by the author in the spring of 2022 and which was apparently inspired by the aforementioned additive and multiplicative equations. With p be an odd prime, we consider the numbers n such that the following congruence equation admits a solution:

$$n \equiv a + b \equiv ab \pmod{p}, \tag{C3}$$

$$n \equiv a - b \equiv ab \pmod{p}. \tag{C4}$$

It is not difficult to prove that a sufficient condition for (C3) to have a solution when $n = 1$ is that p can be written as $x^2 + 3y^2$, and a sufficient condition for (C4) to have a solution is that p can be written as $5x^2 - y^2$.

Let the set of integers

$$S_+ = \{n \in Z_p^* | n \equiv a + b \equiv ab \pmod{p}\},$$

$$S_- = \{n \in Z_p^* | n \equiv a - b \equiv ab \overline{\pmod{p}}\}.$$

Here, Z_p^* denotes the reduced residue system modulo p. We have investigated the congruences modulo p of products and power sums of elements in S_+, the S_-, respectively, with the following results:

$$\Pi_{n \in S_+} n \equiv -2 \pmod{p},$$

$$\sum_{n \in S_+} n^k \equiv \begin{cases} 2^{2s-1} - \frac{1}{2}\binom{2s}{s} \pmod{p}, & \text{if } s \neq 0 \\ \frac{p-1}{2} \pmod{p}, & \text{if } s = 0. \end{cases}$$

Here, $0 \leq s < p - 1$, $s \equiv k \pmod{p-1}$. In particular, when $n = -1$ and -2, using Fermat's little theorem, it follows that

$$\sum_{n \in S_+} \frac{1}{n} \equiv \frac{1}{8} \pmod{p}, \quad p > 3,$$

$$\sum_{n \in S_+} \frac{1}{n^2} \equiv \frac{1}{32} \pmod{p}, \quad p > 5.$$

These are analogues of Wilson's theorem and Wolstenholme's theorem.

Further, let R denote the set of quadratic residues modulo p, N the remaining set of nonquadratic residues modulo p. We prove that

$$|S_+ \cap R| = \frac{1}{4}\left(p - \left(\frac{-1}{p}\right)\right)$$

$$|S_+ \cap N| = \frac{1}{4}\left(p - 2 + \left(\frac{-1}{p}\right)\right),$$

$$\prod_{n \in S_+ \cap R} n \equiv \frac{3}{2} - \frac{5}{2}\left(\frac{-1}{n}\right) \pmod{p}.$$

$$\sum_{n \in S_+ \cap R} n^k \equiv \begin{cases} -\frac{1}{4}\left(\frac{-1}{p}\right) \pmod{p}, & \text{if } s = 0, \\ 2^{2s-1} - \frac{1}{4}\binom{2s}{s} \pmod{p}, & \text{if } 0 < s < \frac{p-1}{2}, \\ 2^{2s-1} - \frac{1}{4}\left(\binom{2s}{s} + 2\binom{2s}{s-\frac{p-1}{2}}\right) \pmod{p}, & \text{if } \frac{p-1}{2} \leq s < p - 1. \end{cases}$$

Replacing S_+ by S_- produces a similar set of four congruences.

Consider the famous Hadamard conjecture in combinatorial mathematics, which says that for n a multiple of 4, there exists a Hadamard matrix of order n, where a Hadamard matrix is a square matrix of order n with elements $a_{ij} = \pm 1$ and mutually orthogonal rows. The main method of constructing Hadamard matrices in the past was by way of a matrix invented by the British mathematician Paley, drawing upon the Legendre symbol for quadratic residues, Since S_+ and S_- like the sets R and N each have $\frac{p-1}{2}$ elements, it may be possible to use them to design and construct a new Hadamard matrix to complete the final proof of Hadamard's conjecture.

As for other applications of the additive and multiplicative equations, we the Erdős-Strauss conjecture and the Schipiński conjecture associated with the famous Egyptian fractions. Originally, these were regarded as similarly difficult, or at least it was difficult to draw any conclusions. However, after decomposing them using the idea of additive and multiplicative equations, we find that one of these two variants of the conjecture is clearly valid, while the other one remains open. In other words, there is some difference in the difficulty of the problem.

About the Author

Cai Tianxin graduated from Shandong University and obtained a PhD in mathematics in 1987 and has been teaching at Zhejiang University in Hangzhou, China, ever since, becoming a professor in 1994. He has also visited dozens of universities including Cambridge University, Princeton University, Brown University, Université Paris Cité, Université Sorbonne, the University of Göttingen, Utrecht University, and the University of Melbourne. He is not only a number theorist but also a poet and writer, who has published a variety of books of literature and popular mathematics and given public lectures in many countries across six continents.

Cai Tianxin's research interests include problems related to perfect numbers, congruence theory modulo integer powers, multiple zeta function and Witten zeta function values, the history of mathematics, and so on. He has proposed and studied many of his own original problems. The topic of additive and multiplicative equations is among these new directions, one that he first began exploring in 2011, when he modified an old problem, originally formulated by Waring, by replacing a sum of k-th powers with a sum of ordinary positive integers, requiring only the k-th power of their product. This produced concise and beautiful conclusions for both of the associated numbers $g(k)$ and $G(k)$, although the challenges in dealing with $G(k)$ remained. Subsequently, he applied this concept of additive and multiplication equations to some classical problems in number theory, including Fermat's last theorem, Euler's conjecture, the representation of integers as sums of squares, congruent numbers, and Egyptian fractions. He also considered figurate primes, additive and multiplicative congruences, and *abcd* equations, leading to other new variations.

According to the subject classification of the American Mathematical Society, there exist separately additive number theory and multiplicative number theory. The contents of the above discussion constitute this new book, *Additive and Multiplicative Number Theory*, which not only details both the progress and challenges of these new problems but also reviews their historical origins and progress toward the corresponding classical version. This is Cai Tianxin's ninth work in English; the others include *A Modern Introduction to Classical Number Theory* and *Perfect Numbers and Fibonacci Sequences*, both also published by World Scientific. His books of collected poetry in English include "Song of the Quiet Life" and "Every Cloud Has Its Own Name", published in South Africa and United States, respectively, and his science popularization works include "Mathematical Legends — from Thales to Erdős" by Springer and "A Brief History of Mathematics — A Promenade through the Civilizations of Our World" by Birkhäuser.

Contents

Chapter 1

The New Waring Problem

> Everything that our minds can comprehend is mutually
> interconnected.
>
> —Leonhard Euler

1.1   Origins of the Problem

Around 1736, Waring was born in Shropshire, west of Birmingham,
which borders Wales to the west and is one of the most sparsely
populated areas in England. Like Newton, Waring was the eldest son
of a farm owner and later entered Maudlin College, Cambridge, as
a reduced fee student (Fig. 1.1). He soon showed exceptional talent
in mathematics, and in 1757 received his bachelor's degree with first
class honors and stayed on to teach; 2 years later he was appointed
to the prestigious Lucasian Chair, a position that Newton had held.

Because Waring was so young and because he had not yet received
his master's degree (which was usually considered necessary), the
president of St. John's College objected to granting him the Lucasian
Chair, but his jurist friend John Wilson supported him. Three years
later, Waring was elected to the Royal Society, from which he with-
drew in 1795, citing on account of his age; he was also a member of
the Academies of Göttingen, Germany, and Bologna, Italy. In 1767,
Waring received a doctorate in medicine and did human dissection,
but his medical career was less successful and very brief.

Figure 1.1.   British mathematician Waring.

In this book *Meditationes Algebraicae (Meditations on Algebra)*, Waring records a conjecture proposed by Wilson, that

if p is any prime number, then $(p-1)! \equiv -1 \,(\mathrm{mod}\ p)$.

In 1801, Wilson's theorem was extended to a general proposition concerning positive integers (Gauss's Theorem) by Carl Friedrich Gauss (1777–1855), a German mathematician of the same age who lived in Berlin.

Also in 1770, Lagrange proved his sum of four squares theorem, which gives a perfect solution to the Waring problem when $k = 2$ (Fig. 1.2). In fact, the ancient Greek mathematician Diophantus of

Figure 1.2.   Lagrange, a descendant of France and Italy.

Alexandria (active around 250 AD) probably already knew this conclusion, judging from examples he gave in his *Arithmetica*. The official authorship of this conjecture is attributed to the French scholar, poet, and theologian Claude-Gaspar Bachet (1581–1638), who was the translator for the 1621 Latin edition of *Arithmetica*.

This Bachet translation was at least the third Latin translation of *Arithmetica*, which he published at his own expense, incorporating some personal observations and notations. In particular, Bachet verified by inspection that all positive integers up to 120 can be expressed as a sum of squares of four integers. Like Euclid's *Geometry*, *Arithmetica* was written in 13 volumes, the first six of which were found in a Greek translation in Venice, Italy, in 1464; an Arabic translation of another four volumes were found only relatively recently in Mashhad, Iran.

The French mathematician Pierre de Fermat (1601–1665), known as the "king of amateur mathematicians", notably owned the Bachet translation of *Arithmetica*, which apparently consisted of only the first six volumes. It is said to have been purchased by the 20-year-old Fermat in Paris in 1621. In the 18th commentary written in the margin of the book, Fermat claims to have proved this proposition (as he famously did for Fermat's Last Theorem, etc.).

In addition to making notes in the margins of his book *Arithmetica*, Fermat wrote to some of his colleagues about his findings. For example, in a letter to fellow mathematician and physicist Gilles Personne de Roberval (1602–1675), Fermat mentioned that the proof of the above results was quite difficult and that he was finally able to overcome these difficulties by using the method of infinite descent, which he himself had invented.

Unfortunately, one has never found a proof of Fermat, and even the Swiss mathematician Leonhard Euler (1707–1783), who later explored each of the problems studied by Fermat, was not able to give a proof. In fact, this 18th commentary ended up constituting the following theorem.

Fermat's polygonal number theorem. *For $n \geq 3$, all natural numbers can be expressed as a sum of no more than n n-gonal numbers.*

The n-gonal number here is the number of black dots arranged in a n-gonal array, as depicted in the figure, and they belong to a general type of figurate number, or number corresponding to the arrangement of a certain shape. When the Pythagoreans studied the concept of number, they liked to depict numbers as small stones on a beach. The small stones can form different geometric figures, and thus a series of figurate numbers is created.

For example, $1, 3, 6, 10, 21, \ldots$ are triangular numbers, i.e., binomial coefficients

$$\frac{m+1}{2} = \frac{m(m+1)}{2}, \quad m \geq 1.$$

In particular, the wooden bottles of bowling and the target balls of snooker are arranged in the form of triangles, and their numbers are 10 and 21, respectively. Also, the quadrangular numbers $1, 4, 9, 16, 25, \ldots$ are square numbers. The general term of an n-gonal number is

$$\frac{m}{2}\{(n-2)m - (n-4)\}, \quad m \geq 1.$$

In his commentary, Fermat mentions that the above proposition shows the mystery and profundity of number theory, and says that he intended to write a book about it, but did not do so. A century later, when Euler read this commentary, he became quite excited and at the same time regretted that Fermat did not leave behind any proofs. One could say that it was Fermat who elevated this ancient mathematical game into a problem. Since then, Euler became Fermat's successor in the field of number theory, and he argued for many of Fermat's assertions, including the famous Fermat–Euler theorem in congruence theory, but he failed to give an answer to the problem of n-gonal numbers.

It was not until 1770 that Lagrange, building on Euler's work, finally proved the case $n = 4$, in other words, the sum of four squares theorem. The proof of $n = 3$ was given by Gauss in 1796, when he was only 19 years old. As for the general case, it was proved by the French mathematician Augustin-Louis Cauchy (1789–1857) in 1813 (Fig. 1.3), the year of Lagrange's death.

For his youthful proof of the case $n = 3$, Gauss wrote it down in his notebook with the note Eureka!, echoing the words of the ancient Greek mathematician Archimedes (287–212 BC) when he discovered the law of floating bodies, and it is clear that Gauss himself took this result very seriously. In fact it is not uncommon among problems in number theory that the case $n = 3$ is more difficult than the case of $n = 4$; this is true also of Fermat's Last Theorem, which

Figure 1.3. Statue of Cauchy, photograph by the author in Ecole Normale Superieure, Paris.

was also proved first for $n = 4$ (Fermat, 1637) and later for $n = 3$ (Euler, 1753). Euler announced this result in a letter to the German mathematician Christian Goldbach (1690–1764) while he was a guest in Berlin.

Gauss first proved that all positive integers of the form $8n + 3$ can be expressed as a sum of squares of three integers. Then, assuming $8n + 3 = a^2 + b^2 + c^2$, it is easy to see that a, b, and c must all be odd, so that $a = 2x - 1, b = 2y - 1, c = 2z - 1$. Then we have an algebraic identity

$$\frac{(2x-1)^2 + (2y-1)^2 + (2z-1)^2 - 3}{8}$$
$$= \frac{x(x-1)}{2} + \frac{y(y-1)}{2} + \frac{z(z-1)}{2}$$

so that

$$n = \frac{x(x-1)}{2} + \frac{y(y-1)}{2} + \frac{z(z-1)}{2} = \binom{x}{2} + \binom{y}{2} + \binom{z}{2}.$$

For the case $n = 4$, the German mathematician Carl Gustav Jacobi (1804–1851) gave a new and powerful proof in 1828. For an arbitrary nonnegative integer n, suppose that it can be expressed as a sum of four squares in some number $a(n)$ of distinct ways; then Jacobi's proof makes use of the fact that the series

$$\sum_{n=0}^{\infty} a(n)e^{2\pi inz}$$

is a modular form of a certain level.

As for the Waring problem itself, progress had been made in the 139 years prior to Hilbert's proof of general existence for certain special cases, such as $k = 3, 4, 5, 6, 7, 8, 10$. After Hilbert proved that the result holds for arbitrary k, this conclusion also became known as the Hilbert–Waring theorem.

After proof of existence $s(k)$ had been obtained, focus shifted to its size, which one considers by setting $g(k)$ to denote the smallest positive integer such that every positive integer admits a presentation as a sum of $g(k)$ kth powers of nonnegative integers.

Since $7 = 2^2 + 3 \times 1^2$ cannot be written as a sum of three squares, it follows from Lagrange's theorem that $g(2) = 4$.

In 1772, Euler's son, J. A. Euler, speculated that

$$g(k) = 2^k + \left[\left(\frac{3}{2}\right)^k\right] - 2.$$

This conjecture was made on the basis of Euler's result, which proved that for $k \geq 2$, the right-hand side of the above equation is a lower bound on $g(k)$. This conjecture has been confirmed to hold with very great accuracy, but it has not yet been proven. In 1990, Kubina and Wunderlich verified that the conjecture holds for $k \leq 471,600,000$. Mahler proved that there are at most finitely many exceptions to k.

In 1909, the German mathematician Wieferich proved that $g(3) = 9$, but his proof was later found to have contained a gap, which was filled by the British-born American mathematician Kempner (1912). In 1986, three mathematicians, Ramachandran Balasubramanian, Drey Françoise Dress and Jean-Marc Deshouillers collaborated to prove that $g(4) = 19$. In 1940, the Indian mathematician Subbayya Pillai (1901–1950) proved that $g(6) = 73$, and in 1964, the Chinese mathematician Chen Jingrun (1933–1996) proved that $g(5) = 37$.

It is worth mentioning that the work of Balasubramaniam and three others utilized and improved on the method of Chen Jingrun. Pillai, who was a generation ahead of Balasubramanian, is considered one of the most important Indian mathematicians after Ramanujan. There is associated with him the famous Pillai conjecture:

For any positive integer k, there exist at most finitely many pairs of positive integer powers (x^p, y^q) that satisfy $x^p - y^q = k$.

In 1946, Hua Luogeng, then a professor at Southwest Associated University, stopped in Calcutta on his way to the Soviet Union from Kunming and had an exchange with Pillai. Four years later, Pillai was invited to the Institute of Advanced Study at Princeton in the United States for a 1-year visit. He planned to go first to Harvard University to attend the International Congress of Mathematicians, but unfortunately he died in a plane crash on the way to Cairo.

On the other hand, it is known from the work of the British mathematicians G. H. Hardy (1877–1947) and J. E. Littlewood (1885–1977) that there is a more essential function attached to this problem than $g(k)$, which is written as $G(k)$, and denotes the smallest positive integer such that all sufficiently large positive integers all can be written as a sum of $G(k)$ kth powers of nonnegative integers. Obviously $G(k) \leq g(k)$ for all k. Determining the value of $G(k)$ is at once a more complex and more meaningful problem. Using basic properties of congruences, it is easy to see that integers congruent to 7 modulo 8 cannot be expressed as a sum of three squares, so $G(2) = 4$.

In 1939, Davenport proved (see Vaughan, 1982, Chapter 6, Section 2) that $G(4) = 16$, using the circle method of Hardy and Littlewood. More precisely, he proved the following conclusion: any sufficiently large $n \not\equiv 0$ or $1 \pmod{16}$ can be expressed as a sum of the fourth powers of fourteen integers. Then writing $n = n - 1 + 1 = n - 2 + 1 + 1$, it follows that any sufficiently large integer can be written as a sum of the fourth powers of sixteen integers, so $G(4) \leq 16$.

On the other hand, the reverse inequality $G(4) \geq 16$ can be proved using elementary methods (see, for example, Hua 1982, Chapter 18, Section 2). Note that $x^4 \equiv 0$ or $1 \pmod{16}$, so any positive integer of the form $16m + 15$ requires at least fifteen fourth powers; so $G(4) \geq 15$. Now, if $16 \cdot n = x_1^4 + \cdots + x_{15}^4$, then $2 \mid (x_1, \ldots, x_{15})$, so $16 = y_1^4 + \cdots + y_{15}^4$ where each $y_i = \frac{x_i}{2}(1 \leq i \leq 15)$. Then since 31 cannot be expressed as a sum of fewer than 16 fourth powers, it follows that $16 \cdot 31$ cannot be expressed as a sum of 15 fourth powers, and continuing on in this way we generate an infinite sequence of numbers none of which can be expressed as a sum of 15 fourth powers; so so $G(4) \geq 16$.

This is a surprising result. For beyond this, one can only determine the range of $G(k)$, e.g., $4 \leq G(3) \leq 7$, although it has been conjectured that $G(3) = 4$. So far, the largest known integer that cannot be written as a sum of four non-negative integer cubes is $7, 373, 170, 279, 850$. Little is known for $k > 4$; we have $6 \leq G(5) \leq 17, 9 \leq G(6) \leq 24$. In 1996, Hongze Li (see Li, 1996) showed that $G(16) \leq 111$.

On the other hand, Hardy and Littlewood conjectured that for $k \neq 2^a, a > 1, G(k) \leq 2k+1$. At present, the upper bound estimate of $G(3)$ is the corroborating evidence for this conjecture. It is a pity that the Waring problem, one of the central problems of number theory, was not included in Hilbert's 23 mathematical problems for the future in 1900. Perhaps in 1900, Hilbert had not yet been exposed to the problem, since once the Hilbert–Waring theorem became available, it began to attract more and more number theorists. Nevertheless, progress on the Waring problem has been very slow over a long period of time.

1.2   Variants of the Waring problem

In April 2011, the authors proposed a class of variants of the Waring problem. Let k and s be positive integers and consider the diophantine equation

$$n = x_1 + x_2 + \cdots + x_s$$

subject to the condition

$$x_1 x_2 \cdots x_s = x^k$$

By the Hilbert–Waring Theorem, there exists $s = s'(k) \leq s(k)$ such that for any positive integer n, it can be expressed as a sum of no more than s positive integers whose product is a kth power. Denote by $g'(k)$ (respectively $G'(k)$) that smallest positive integer s such that any (respectively any sufficiently large) positive integer n can be expressed as a sum of no more than $g'(k)(G'(k))$ positive integers the product of which is a kth power. It is obvious that

$g'(k) \leq g(k), G'(k) \leq G(k)$. This problem has been studied by the author and Deyi Chen (see Cai-Chen, 2013), who proved that

Theorem 1.1. *For any positive integer* $k, g'(k) = 2k - 1$.

Proof. When $k = 1$ obviously, let $k > 1$, $n \equiv i \pmod{k}$, $0 \leq i \leq k - 1$, then we have

$$n = \underbrace{1 + \cdots + 1}_{i} + \underbrace{\frac{n-i}{k} + \cdots + \frac{n-i}{k}}_{k}.$$

Therefore, $g'(k) \leq 2k - 1$.

On the other hand, $2k - 1$ cannot be expressed as the sum of less than $2k - 1$ positive integers such that the product is a kth power. Otherwise, there must be a positive integer greater than 1.

Let the smallest of its prime factors be q. Then $2k - 1 \geq q^{\beta_1} + \cdots + q^{\beta_r}$, where $\beta_1 \geq 1, \ldots, \beta_r \geq 1, \beta_1 + \cdots + \beta_r$ is a multiple of k. Thus

$$2k - 1 \geq \underbrace{q + \cdots + q}_{k} = qk \geq 2k.$$

This is a contradiction, so $g'(k) \geq 2k - 1$. Theorem 1.1 is proved. $\square$

For $G'(k)$, we obtain multiple lower bound estimates. For example.

Theorem 1.2. *For any prime number* $p, G'(p) \leq p + 1$.

Proof. Let $n \equiv i \pmod{p}, 0 \leq i \leq p - 1$. If $i = 0$, set $n = mp$; when $n > p^p$ we have

$$n = m - p^{p-1} + \cdots + m - p^{p-1} + p^p.$$

If $0 < i \leq p - 1$ then by Fermat's little theorem $n \equiv i \equiv i^p \pmod{p}$. When $n > p^p$, we have

$$n = i^p + \underbrace{\frac{n - i^p}{p} + \cdots + \frac{n - i^p}{p}}_{p}.$$

Theorem 1.2 is proved. $\square$

In the spring of 2017, in the course of an introductory number theory course delivered by the author, Junwei Hu, a third-year finance major, improved upon Theorem 1.2. He proved (cf. Cai, 2021) the following theorem.

Theorem 1.3. *If m is a cyclic number (a number m satisfying $(m, \varphi(m)) = 1$), then $G'(m) \leq m + 1$.*

Obviously, the cyclic numbers include all prime numbers, and any cyclic number must be squarefree. Let $n = km + r$, $0 \leq r < m \leq r < m$. Let $d = (r, m)$. Since m has no square factor, $(r, \frac{m}{d}) = 1$, and by Euler's theorem

$$r^{\phi\left(\frac{m}{d}\right)} \equiv 1 \left(\bmod \frac{m}{d}\right).$$

Now for any integer B, it is always the case that

$$r^{\phi\left(\frac{m}{d}\right)B} \equiv 1 \left(\bmod \frac{m}{d}\right).$$

Since d divides r, we have

$$r^{\phi\left(\frac{m}{d}\right)B+1} \equiv r \,(\bmod\, m). \tag{1.1}$$

Moreover from $(m, \phi(m)) = 1$, we know that also $\left(m, \phi\left(\frac{m}{d}\right)\right) = 1$, and therefore there exist positive integers A_r, B_r satisfying

$$\phi\left(\frac{m}{d}\right) B_r + 1 = A_r m.$$

Applying (1.1) to the above equation, we get

$$r \equiv r^{A_r m} \,(\bmod\, m).$$

Take $C_r = \frac{r^{A_r m} - r}{m}$; then C_r is a positive integer. Setting $C = \max_{0 \leq r < m} C_r$, we find that when $n > m(C + 1)$,

$$k = \left[\frac{n}{m}\right] \geq [C + 1] = C + 1 > C \geq C_r.$$

So

$$n = k - C_r + \cdots + k - C_r k - C_r + \cdots + k - C_{r+m} C_r + r$$

$$= k - C_r + \cdots + k - C_{r+} r^{A_r m_r}$$

and the product of the terms is $(k - C_r)^m (r^{A_r})^m (k - C_r)^m (r^{A_r})^m$. Therefore, $G'(m) \leq m + l$, and Theorem 1.3 is proved.

In particular, we would like to point out that Carmichael numbers are cyclic numbers. A Carmichael number is a composite number n such that for any $a > 1$ with $(a, n) = 1$,

$$a^n \equiv a \pmod{n}.$$

According to Korselt's criterion, any Carmichael number n must be squarefree and such that for every prime $p | n, p - 1 | n - 1$.

For any two prime factors p and q of n, it is easy to see that $(n - 1, q) = 1$, so $(p - 1, q) = 1$, and thus we have $(n, \varphi(n)) = 1$.

Example 1.1. 561 is the smallest Carmichael number.

We verify only that 561 is a Carmichael number. $561 = 3 \times 11 \times 17$, if $(a, 561) = 1$, then $(a, 3) = 1, (a, 11) = 1, (a, 17) = 1$. By Fermat's little theorem, $a^2 \equiv 1 \pmod 3$ $a^{10} \equiv 1 \pmod{11}$ $a^{16} \equiv 1 \pmod{17}$. Considering $[2, 10, 16] = 80$, we have $a^{80} \equiv 1 \pmod 3$ $a^{80} \equiv 1 \pmod{11}$ $a^{80} \equiv 1 \pmod{17}$, and so $a^{80} \equiv 1 \pmod{561}$. Then by the fact that 560 is a multiple of 80 , we know that 561 is a pseudo-prime number with base a, and thus 561 is a Carmichael number.

Example 1.2. Let $n = p_1 p_2 \ldots p_k$, the primes p_i be all distinct with $k > 2$ such that for any i, $p_i - 1 \mid n - 1$. Then n is a Carmichael number.

Some of the smaller Carmichael numbers are

$$1105 = 5 \times 13 \times 17, 1729 = 7 \times 13 \times 19, 2465$$
$$= 5 \times 17 \times 29, 2821 = 7 \times 13 \times 31.$$

The third smallest Carmichael number, 1729, also known as the Hardy–Ramanujan number, after a story recalled by the British mathematician Hardy, who once visited the ailing Indian mathematician Ramanujan via a taxi with license number 1729, a number he thought to have no special significance. Instead, Ramanujan told Hardy that it was the smallest positive integer that could be expressed in two ways as a sum of two integer cubes, i.e., $1729 = 1^3 + 12^3 = 9^3 + 10^3$.

Question 1.1. Is it true that $G'(m) \leq m + 1$ for any squarefree number m?

Theorem 1.4. *For any positive integer* $k, G'(k) \leq k + 2$.

Proof. Proof For any positive integer n, let $n = km + r$, where $0 \le r \le k - 1$. If $r = 0$, then when $n > 2k^{2k}$, we have $m = \frac{n}{k} > 2k^{2k-1}$, so

$$n = km = \underbrace{(m - 2k^{2k-1}) + \cdots + (m - 2k^{2k-1})}_{k} + k^{2k} + k^{2k}.$$

If $0 < r \le k - 1$, then when $n > k^{2k-1} + k$, we have $m = \frac{n-r}{k} > \frac{k^{2k-1}}{k} > k^{k-1}r^{k-1}$, so

$$n = km + r = (m - k^{k-1}r^{k-1}) + (m - k^{k-1}r^{k-1}) + k^k r^{k-1} + r.$$

Theorem 1.4 is proved.

Notice that when $k > 1$, a prime number congruent to 3 modulo 4 cannot be expressed as the sum of two integers to the kth power. By Theorems 1.1 and 1.2, $G'(2) = 3, 3 \le G'(3) \le 4, 3 \le G'(5) \le 6$; by Theorems 1.1 and 1.4, $3 \le G'(4) \le 6$.

To better estimate $G'(3)$, we set

$$S = \{n \mid n = x + y + z, xyz = m^3, x, y, z \in Z_+\}.$$

Obviously, whenever n belongs to S, then also the multiples of n belong to S. Therefore, we need only consider the case where n is a prime number. We have the following results. □

Theorem 1.5. *If p is a prime number, $p \equiv 1 \pmod{3}$ or $p \equiv \pm 1 \pmod{8}$, then $p \in S$.*

Theorem 1.6. *Suppose p is a prime number, $p \equiv 5 \pmod{24}$, and there exists a pair of positive integers (x, y) satisfying*

$$p = 6x^2 - y^2, \tag{1.2}$$

with

$$\frac{x}{y} \in \left(\frac{7}{17}, \frac{19}{46}\right) \cup \left(\frac{11}{26}, \frac{3}{7}\right) \cup \left(\frac{9}{19}, \frac{7}{13}\right) \cup \left(1, \frac{3}{2}\right), \tag{1.3}$$

then $p \in S$.

After a computer check, only $2, 5, 11 \notin S$ *among all prime numbers up to* 10000. *Notice that*

$$2^6 = 7 + 8 + 49, \quad 7 \times 8 \times 49 = 14^3,$$
$$5^2 = 3 + 4 + 18, \quad 3 \times 4 \times 18 = 6^3,$$
$$11^2 = 1 + 45 + 75, \quad 1 \times 45 \times 75 = 15^3.$$

It follows that there are only finitely many integers of the form $2^\alpha 5^\beta 11^\gamma$ *that do not belong to* S. *On the basis of the above calculation (a similar operation can be performed for* $k \geq 4$*), we make the following conjecture:*

Conjecture 1.1. *Every positive integer except* $1, 2, 4, 5, 8, 11, 16,$ $22, 32, 44, 88, 176$ *can be expressed as a sum of three positive integers whose product is the cube of an integer. In particular,* $G'(3) = 3$.

Conjecture 1.2. *Every positive integer except* $1, 2, 3, 5, 6, 7, 11, 13,$ $14, 15, 17, 22, 23$ *can be expressed as a sum of four positive integers whose product is the fourth power of an integer.*

In particular, $G'(4) = 4$.

Although we have not been able to prove the above conjectures, we proved that the conclusions of these two conjectures hold when one or both of the positive integers are allowed to range over all integers.

Theorem 1.7. *Let* $n > 1$, *there exist positive integers* (x, z), *integers* $(y, m), gcd(x, y, z) < n$ *such that*

$$n = x + y + z$$

with

$$xyz = w^3.$$

Theorem 1.8. *Let* $n > 1$, *there exist positive integers* (x, z, m), *negative integers* (y, w), *and* $gcd(x, y, z, w) < n$ *such that*

$$n = x + y + z + w$$

with

$$xyzw = m^4.$$

1.3   Proof of Theorems

To prove Theorems 1.5–1.7, we need the following lemmas:

Lemma 1.1. *Let p be a prime number, $p \equiv \pm 1 (\mathrm{mod}\,8)$. Then there exist positive integers (x, y) that satisfy*

$$x^2 - 2y^2 = p, \tag{1.4}$$

and $x > 2y$.

Lemma 1.2. *Let p be a prime number, $p \equiv 5\ 11 (\mathrm{mod}\,24) \equiv 5\ 11 (\mathrm{mod}\,24)$, then there exist positive integers (x, z) and integers $(y, m), gcd(x, y, z) = 1$, such that*

$$p = x + y + z$$

and

$$xyz = m^3.$$

To prove Lemmas 1.1 and 1.2, we use some basic properties of quadratic forms (see Alaca-Williams, 2004, or Weisstein, 1995), for any prime p, if $p \equiv \pm 1 (\mathrm{mod}\,8)$, then we have a presentation as $p = x^2 - 2y^2$; if $p \equiv 5 (\mathrm{mod}\,24)$, then we have $p = 6x^2 - y^2$; and if $p \equiv 11 (\mathrm{mod}\,24)$, then $p \equiv 3 (\mathrm{mod}\,8), p = x^2 + 2y^2 x^2 + 2y^2$.

Proof of Lemma 1.1. It is easy to see that $(u, v) = (3, 2)$ is a fundamental solution of the indefinite equation $x^2 - 2y^2 = 1$; moreover if $x + y\sqrt{2}$ is the fundamental solution of (1.4), then by the theory of quadratic forms (see Nagell, 1951, Theorem 108);

$$0 \le y \le \frac{v\sqrt{p}}{\sqrt{2(u+1)}} = \sqrt{\frac{p}{2}}. \tag{1.5}$$

Combining (1.4) and (1.5), we have

$$\left(\frac{x}{y}\right)^2 = 2 + \frac{p}{y^2} \ge 2 + \frac{p}{\left(\sqrt{\frac{p}{2}}\right)^2} = 4$$

therefore

$$\frac{x}{y} \ge 2.$$

Since $x \ne 2y$, the inequality is strict, and Lemma 1.1 is proved.

Proof of Lemma 1.2. First consider $p \equiv 5 \pmod{24}$ and make the following transformations:

$$\begin{cases} x = 3a + b \\ y = 2a + b, \end{cases} \quad \begin{cases} x = 7a + b \\ y = 13a + 2b, \end{cases} \quad \begin{cases} x = a + 9b \\ y = 2a + 19b, \end{cases}$$

$$\begin{cases} x = 3a + 11b \\ y = 7a + 26b, \end{cases} \quad \begin{cases} x = 19a + 7b \\ y = 46a + 17b. \end{cases} \tag{1.6}$$

Substituting into (1.2), we get in turn

$$p = 50a^2 + 32ab + 5b^2, \quad p = 125a^2 + 32ab + 2b^2,$$
$$p = 2a^2 + 32ab + 125b^2,$$
$$p = 5a^2 + 32ab + 50b^2, \quad p = 50a^2 + 32ab + 5b^2.$$

The inverse transformations are

$$\begin{cases} a = x - y \\ b = 3y - 2x, \end{cases} \begin{cases} a = 2x - y \\ b = 7y - 13x, \end{cases} \begin{cases} a = 19x - 9y \\ b = y - 2x, \end{cases} \begin{cases} a = 26x - 11y \\ b = 3y - 7x, \end{cases}$$

$$\begin{cases} a = 17x - 7y \\ b = 19y - 46x \end{cases} \tag{1.7}$$

and the nature of the reciprocals is obvious.

If $p \equiv 11 \pmod{24}$, we have instead the following transformations:

$$\begin{cases} x = 5a \\ y = a + b, \end{cases} \quad \begin{cases} x = 2a + b \\ y = 5a, \end{cases} \quad \begin{cases} x = a + 4b \\ y = -a + b, \end{cases}$$

$$\begin{cases} x = a + 4b \\ y = a - b \end{cases}, \quad \begin{cases} x = 2a - b \\ y = a + 2b \end{cases}, \quad \begin{cases} x = a - 2b \\ y = 2a + b. \end{cases}$$

Substitute $p = x^2 + 2y^2$, in order to obtain

$$p = 27a^2 + 4ab + 2b^2, \quad p = 54a^2 + 4ab + b^2, \quad p = 3a^2 + 4ab + 18b^2.$$
$$p = 3a^2 + 4ab + 18b^2, \quad p = 6a^2 + 4ab + 9b^2, \quad p = 9a^2 + 4ab + 6b^2.$$

Their inverse transformations are

$$\begin{cases} a = \frac{x}{5} \\ b = y - \frac{x}{5}, \end{cases} \quad \begin{cases} a = \frac{y}{5} \\ b = x - \frac{2y}{5}, \end{cases} \quad \begin{cases} a = \frac{x+y}{5} - y \\ b = \frac{x+y}{5}, \end{cases}$$

$$\begin{cases} a = \frac{x-y}{5} + y \\ b = \frac{x-y}{5}, \end{cases} \quad \begin{cases} a = \frac{2x+y}{5} \\ b = \frac{2(2x+y)}{5} - x, \end{cases} \quad \begin{cases} a = \frac{x+2y}{5} \\ b = y - \frac{2(x+2y)}{5}. \end{cases}$$

The following proof shows that one of the above six inverse transformations must have both a and b as integers. To do so, we only need to prove that one of the following 6 numbers is an integer:

$$\frac{x}{5}, \frac{y}{5}, \frac{x+y}{5}, \frac{x-y}{5}, \frac{2x+y}{5}, \frac{x+2y}{5}.$$

This point can be derived from the following congruence relation using Fermat's little theorem

$$xy(x+y)(x-y)(2x+y)(x+2y) = 5^{x^4}y^2 + 2x^5y - 5x^2y^4 - 2xy^5$$
$$\equiv 2xy\left(x^4 - y^4\right) \equiv 0(\bmod 5).$$

It is obvious that the condition of mutual coprimality is satisfied. Lemma 1.2 is proved.

Remark 1.1. *The proof of the first half of Lemma 1.2 requires only one of the transformations in (1.6), we have included five here for the sake of the proof of Theorem 1.6.*

Proof of Theorem 1.5. If $p \equiv 1(\bmod 3)$, then there exist two positive integers $x, y, (x, y) = 1$ such that

$$p = x^2 + xy + y^2,$$

and if $p \equiv \pm 1(\bmod 8)$, then by Lemma 1.1, there exist two positive integers $x, y, x > 2y, (x, y) = 1$ such that

$$p = x^2 - 2y^2.$$

Make the substitutions
$$\begin{cases} x = 2a + b \\ y = a, \end{cases}$$
which give
$$p = (2a + b)^2 - 2a^2 = 2a^2 + 4ab + b^2$$
$p \in S$. Theorem 1.5 is proved.

Proof of Theorem 1.6. In the proof of Lemma 1.2, if the inverse transformation (1.7) satisfies one of the following conditions:
$$\begin{cases} x - y > 0 \\ 3y - 2x > 0, \end{cases} \begin{cases} 2x - y > 0 \\ 7y - 13x > 0, \end{cases} \begin{cases} 19x - 9y > 0 \\ y - 2x > 0, \end{cases} \begin{cases} 26x - 11y > 0 \\ 3y - 7x > 0, \end{cases}$$
$$\begin{cases} 17x - 7y > 0 \\ 19y - 46x > 0. \end{cases}$$

Then we have $a > 0, b > 0, p \in S$. And the above condition is equivalent to (1.3), Theorem 1.6 is proved.

Proof of Theorem 1.7. Let $n > 1$ and there exists a prime p such that $n = pn_1$ with n_1 a positive integer. If $p = 2$, then $2 = 3 - 9 + 8$, and $gcd(3, -9, 8) = 1$. If p is an odd prime, it is easy to see that it satisfies one of the following four congruences:
$$p \equiv 1 \pmod 3, \quad p \equiv \pm 1 \pmod 8,$$
$$p \equiv 5 \pmod 8, \quad p \equiv 11 \pmod{24}.$$

By Theorem 1.5 and Lemma 1.2, there exist positive integers (x, z) and integers (y, m), with $gcd(x, y, z) = 1$ such that
$$p = x + y + z$$
and
$$xyz = w^3.$$
So
$$n = pn_1 = n_1 x + n_1 y + n_1 z$$
where
$$n_1 x \cdot n_1 y \cdot n_1 z = (n_1 m)^3$$
$$(n_1 x, n_1 y, n_1 z) = n_1(x, y, z) = n_1 < n.$$
Theorem 1.7 is proved.

Proof of Theorem 1.8. It is easy to see that we only need to consider the case where n is a prime p. If $p = 2$, we have $2 = 6 - 3 + 8 - 9$, and

$$6 \times (-3) \times 8 \times (-9) = 6^4.$$

If p is an odd prime, it is easy to see that it satisfies one of the following four congruent equations:

$$p \equiv 1 \pmod 3, \ p \equiv \pm 1 \pmod 8, \ p \equiv 3 \pmod 8, \ p \equiv 5 \pmod{24}.$$

Using the conclusions in Weisstein 1, we have

$$p = x^2 + 3y^2 = 4x^2 - 3x^2 + 12y^2 - 9y^2,$$
$$p = x^2 + 2y^2 = 2x^2 - x^2 + 4y^2 - 2y^2,$$
$$p = x^2 - 2y^2 = 2x^2 - x^2 + 2y^2 - 4y^2,$$
$$p = 2x^2 - 3y^2 = 6x^2 - 4x^2 + 6y^2 - 9y^2.$$

The four numbers contained in each expression are relatively prime, and their products are, in $(6xy)^4, (4xy)^4, (4xy)^4, (6xy)^4$ $(6xy)^4, (4xy)^4, (4xy)^4, (6xy)^4$ order.

Theorem 1.8 is proved.

Example 1.3. $3 = 1 + 1 + 1, 6 = 2 + 2 + 2, 7 = 1 + 2 + 4,$ $9 = 3 + 3 + 3, 10 = 1 + 1 + 8, 12 = 4 + 4 + 4, 13 = 1 + 3 + 9, \ldots$

Table 1.1. G′(3) data table.

p	x	y	z	m	p	x	y	y	m
3	1	1	1	1	47	2	20	25	10
7	1	2	4	2	53	3	18	32	12
13	1	3	9	3	59	1	4	54	6
17	1	8	8	4	61	2	27	32	12
19	4	6	9	6	67	4	14	49	14
23	2	9	12	6	71	1	20	50	10
29	1	1	27	3	73	1	8	64	8
31	1	5	25	5	79	2	28	49	14
37	9	12	16	12	83	3	8	72	12
41	2	3	36	6	89	5	9	75	15
43	1	6	36	6	97	1	24	72	12

Table 1.2. G$'$(4) data table.

p	x	y	z	w	m	p	x	y	z	w	m
19	1	1	1	16	2	59	4	12	16	27	12
29	2	6	9	12	6	61	1	3	9	48	6
31	3	3	9	16	6	67	1	16	25	25	10
37	1	6	12	18	6	71	3	8	24	36	12
41	1	4	18	18	6	73	2	12	27	32	12
43	2	2	12	27	6	79	2	9	32	36	12
47	1	3	16	27	6	83	2	9	24	48	12
53	8	9	12	24	12	89	2	2	4	81	6
						97	1	12	36	48	12

In general, for prime $p < 100$, the equation $p = x + y + z$, $xyz = m^3$ has solutions given by Table 1.2, where m is the minimum value.

Example 1.4. $4 = 1 + 1 + 1 + 1, 8 = 2 + 2 + 2 + 2, 9 = 1 + 2 + 2 + 4$, $10 = 1 + 1 + 4 + 4, 12 = 3 + 3 + 3 + 3, \ldots$

In general, for prime $p < 100$, the equation $p = x + y + z + w$, $xyzw = m^4$ has the following table of solutions, where m is the minimum value.

The New Waring problem leaves many open questions. In general, for any positive integer $m \geq 3$, we conjecture that $G'(m) = m$. Zhiwei Sun has told the author that he has checked $n = 5$ and 6 with a computer and he thinks this result should be correct.

1.4 Number Sets with Identical Sums and Products

In this section, we investigate a problem posed and studied by the Polish mathematician Andrzej Schinzel. Although it is not strictly a Waring-type problem as the previous problems have been, it involves a similar combination of additive and multiplicative equations, justifying its inclusion in this chapter.

In 1956, M.W. Mnich asked (see Chakraborty, 2007) whether the Diophantine equation

$$x_1 + x_2 + x_3 = x_1 x_2 x_3 = 1$$

admits rational solutions. L.E. Mordell noticed (see Mordell, 1955) that it was related to the equation he had been studying, in particular that this equation has rational solutions if and only if

$$(x_1 + x_2 + x_3)^3 = x_1 x_2 x_3$$

or

$$x_1^3 + x_2^3 + x_3^3 = x_1 x_2 x_3$$

has rational solutions satisfying $x_1 x_2 x_3 \neq 0$. In fact the necessity of this condition is obvious, and the sufficiency is not difficult to prove either.

In 1960, J. Cassels proved (see Cassels, 1960) that this equation has no rational solutions, using the arithmetic of cubic fields. Two years later, G. Sansone and Cassels gave (see Sansone-Cassels, 1962) an elementary proof.

In 1996, Schinzel (see Schinzel, 1996) proved the following theorem.

Theorem 1.9. *The equations*

$$x_1 + x_2 + x_3 = x_1 x_2 x_3 = 6$$

has infinitely many solutions in positive rational numbers.

In 2013, we generalized this result to the case of any $n(\geq 3)$ numbers. Specifically, we proved (*cf. Cai-Zhang, 2013*) that

Theorem 1.10. *The equations*

$$x_1 + x_2 + \cdots + x_3 = x_1 x_2 \ldots x_3 = 2n \qquad (1.8)$$

has infinitely many solutions in positive rational numbers.

To prove Theorems 1.9 and 1.10, we need the following two lemmas, which are well-known theorems in the theory of elliptic curves (see Silverman-Tate, 2004, p. 56 and Skolem, 1950, p. 78, respectively).

Lemma 1.3 (Nagell–Lutz Theorem). *Let Δ be the discriminant of an elliptic curve; if a (x, y) is a rational point on the curve with x and y both integers and either $y = 0$, or $y \mid \Delta$, then the point is of finite order; otherwise, it is of infinite order.*

Lemma 1.4 (Poincaré–Hurwitz Theorem). *If an elliptic curve has infinitely many rational points, then there are infinitely many rational points in the neighborhood of any rational point.*

Proof of Theorem 1.9 (Schinzel). Consider the elliptic curve

$$f(x) = x^3 - 9x + 9 = y^2. \tag{1.9}$$

It has the solution $(x, y) = (7, 17)$ satisfying the conditions of the Nagell–Lutz theorem $17 \nmid \Delta = -729$, where l denotes integer division and Δ is the discriminant of $f(x)$. More precisely, if we set $f(x) = x^3 + ax + b$, $\Delta = 4a^3 + 27b^3$. By the Nagell–Lutz theorem, equation (1.9) has infinitely many rational solutions. Then by the Poincaré–Hurwitz theorem, there are infinitely many rational solutions in any neighborhood of each solution.

Since $(x, y) = (0, 3)$ is a solution of (1.9) and satisfies $|y| < 6 - 3x$.

Therefore, (1.9) has infinitely many solutions (x, y) with $x < 3$.

Let

$$x_1 = \frac{6}{3 - x}, \quad x_2 = \frac{6 - 3x + y}{3 - x}, \quad x_3 = \frac{6 - 3x - y}{3 - x}.$$

Then there is $x_j > 0 (1 \le j \le 3)$, satisfying

$$x_1 + x_2 + x_3 = 6.$$

$$x_1 x_2 x_3 = \frac{6\{(6 - 3x)^2 - y^2\}}{(3 - x)^3} = \frac{6\{(6 - 3x)^2 - f(x)\}}{(3 - x)^3} = 6.$$

Taking different pairs (x, y) generates different triplets (x_1, x_2, x_3). Theorem 1.9 is proved.

Proof of Theorem 1.10. In light of Schinzel's result, we may assume that $n \ge 4$. Taking $x_1 = x_2 = \cdots = x_{n-3} = 1$, (1.8) becomes

$$\begin{cases} x_{n-2} + x_{n-1} + x_n = n + 3 \\ x_{n-2} x_{n-1} x_n = 2n. \end{cases}$$

Eliminate x_n to get

$$x_{n-2}^2 x_{n-1} + x_{n-2} x_{n-1}^2 - (n + 3) x_{n-2} x_{n-1} + 2n = 0$$

or

$$\left(\frac{x_{n-2}}{x_{n-1}}\right)^2 + \frac{x_{n-2}}{x_{n-1}} - (n + 3)\frac{x_{n-2}}{x_{n-1}}\frac{1}{x_{n-1}} + 2n \left(\frac{1}{x_{n-1}}\right)^3 = 0.$$

Set

$$u = \frac{x_{n-2}}{x_{n-1}}, \quad v = \frac{1}{x_{n-1}},$$

so that

$$u^2 + u - (n+3)uv + 2nv^3 = 0.$$

Then we have

$$x = -72nv + 3(n+3)^2, \; y = 216n\{2u + 1 - (n+3)v\}. \quad (1.10)$$

Which gives the elliptic curve (see Figure 1.4 for $n = 4$)

$$E_n : y^2 = x^3 + a_n x + b_n$$

where

$$a_n = -27(n+3)\left(n^3 + 9n^2 - 21n + 27\right),$$

$$b_n = 54n^6 + 972n^5 + 3402n^4 - 5832n^3 + 7290n^2 - 26244n + 39366.$$

This is an elliptic curve over rational numbers; we consider rational points of E_n.

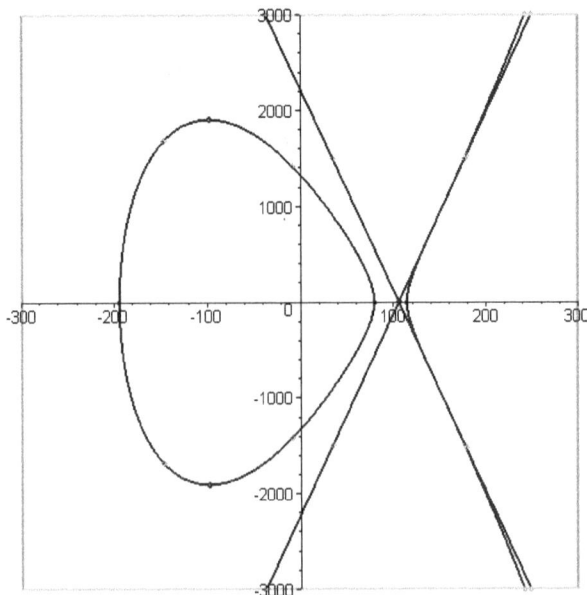

Figure 1.4. Elliptic Curve E4.

First, the discriminant of E_n is

$$\Delta = 2^{15}3^{12}n^3(n^3 + 9n^2 - 27n + 27).$$

When $n \geq 4, \Delta > 0$, so that E_n is non-singular, or in other words, the equation

$$x^3 + a_nx + b_n = 0$$

has three different real roots $x_1(n), x_2(n), x_3(n)$; we can assume that $x_1(n) < x_2(n) < x_3(n)$. From the relationship between the roots and the coefficients, it follows that

$$\begin{cases} x_1(n) + x_2(n) + x_3(n) = 0 \\ x_1(n)x_2(n)x_3(n) = -f(n) \end{cases}$$

where $f(n)$ is the constant term in E_n. It is easy to see that $f(n) > 0$ when $n \geq 4$. Therefore, we have

$$x_1(n) < 0 < x_2(n) < x_3(n).$$

It is not difficult to verify that $P = (3(n + 3)^2, 216n)$, $Q = (3(n - 3)^2, 108n(n - 1))$ and $R = (3(n + 3)^2 - 72n, 216n(n - 2))$ lie on the elliptic curve E_n. Using the group law for the elliptic curve we get

$$[2]P = O, P + Q + R = O$$

$$[2]R = \left(\frac{3\left(n^4 + 2n^3 + 13n^2 - 36n + 36\right)}{(n - 2)^2}, -\frac{216(2n^3 - 6n^2 + 7n - 2)}{(n - 2)^3} \right)$$

$$[3]R = \left(\frac{3\left(n^6 - 36n^4 + 126n^3 - 180n^2 + 108n - 15\right)}{(n^2 - 3n + 3)^2}, \right.$$

$$\left. \frac{108(n - 1)(n - 2)(7n^4 - 33n^3 + 67n^2 - 66n + 28)}{(n^2 - 3n + 3)^3} \right)$$

Here O denotes the point at infinity and $[m]$ indicates the operation

$$[m]P = P + \cdots + P(m \text{ terms})$$

It follows that P is a point of order 2 and P, Q and R lie on the same line.

To prove that there are infinitely many rational points of E_n, it is enough to find a point for which the x-coordinate is not an integer. Using $(n^2 - 3n + 3)^2$ to remove the numerator of the x-coordinate of $[3]R$, we obtain the remainder term as

$$r = -36(3n^3 - 12n^2 + 18n - 10).$$

When $n \geq 4, r \neq 0$, so the x-coordinate of $[3]R$ is not polynomial. For $4 \leq n \leq 109$, it is easy to check that $r/(n^2 - 3n + 3)^2$ is not an integer; when $n > 109$, it is not 0 and its absolute value is less than 1. That is, when $n \geq 4$, none of them is an integer. Therefore, by the Nagell–Lutz theorem, the order is infinite, i.e., for each n, E_n has infinitely many rational points.

Using (1.10), we have the inverse transformation

$$u = \frac{y - 3xn - 9x + 9n^3 + 81n^2 + 27n + 243}{432n},$$

$$v = \frac{3(n+3)^3 - x}{72n}.$$

This leads to

$$x_{n-2} = \frac{y - 3xn - 9x + 9n^3 + 81n^2 + 27n + 243}{6(-x + 3n^2 + 18n + 27)},$$

$$x_{n-1} = \frac{72n}{3(n+3)^2 - x},$$

$$x_n = \frac{-y - 3xn - 9x + 9n^3 + 81n^2 + 27n + 243}{6(-x + 3n^2 + 18n + 27)}.$$

Therefore $(x_1, \ldots, x_{n-3}, x_{n-2}, x_{n-1}, x_n) = (1, \ldots, 1 \cdot x_{n-2}, x_{n-1}, x_n)$ is the solution of (1.8)

Given $x_j > 0, 1 \leq j \leq n$, we have the following conditions:

$$x < 3(n+3)^2, \quad |y| < -3xn - 9x + 9n^3 + 81n^2 + 27n + 243.$$

From the equation $|y| = -3xn - 9x + 9n^3 + 81n^2 + 27n + 243$ the image of the equation can be seen when

$$x < \frac{3(n^3 + 9n^2 + 3n + 27)}{n + 3}$$

when the above condition is satisfied. This is because, when $n \geq 4$, the

$$\frac{3(n^3 + 9n^2 + 3n + 27)}{n + 3} - 3(n + 3)^2 = -\frac{72n}{n + 3} < 0$$

and

$$|y| = g\left(\frac{3(n^3 + 9n^2 + 3n + 27)}{n + 3}\right) = 0.$$

Here $g(x) = -3xn - 9x + 9n^3 + 81n^2 + 27n + 243$.

Consider the equation $|y| = -3xn - 9x + 9n^3 + 81n^2 + 27n + 243$ of the image of the ellipse curve E_n at the point $P = (3(n + 3)^2, 216n)$ and the point $-P = (3(n + 3)^2, -216n)$ with two tangents which intersect at the point

$$\left(\frac{3\left(n^8 + 9n^2 + 3n + 27\right)}{n + 3}, 0\right).$$

It is easy to see that

$$x_2(n) < \frac{3\left(n^3 + 9n^2 + 3n + 27\right)}{n + 3}$$

$$< x_3(n).$$

According to the Poincaré–Hurwitz theorem, infinitely many rational points can be obtained by finding a rational point that fits the preceding condition. From our calculations, the point $[3]R$ meets this condition when $n \geq 4$, i.e.,

$$|y| < -3xn - 9x + 9n^3 + 81n^2 + 27n + 243. \qquad (1.11)$$

This is because the x-coordinate of $[3]R$ is less than the x-coordinate of the intersection of the two tangents of P and $-P$, i.e.,

$$\frac{3(n^6 - 36n^4 + 126n^8 - 180n^2 + 108n - 15)}{(n^2 - 3n + 3)^2} - \frac{3(n^3 + 9n^2 + 3n + 27)}{n + 3}$$

$$= \frac{-36(3n^2 - 7n + 6)(3n^2 - 6n + 4)}{(n^2 - 3n + 3)^2(n + 3)} < 0$$

Therefore, there are infinitely many rational points on the elliptic curve E_n satisfying the condition (1.11), and therefore, there exist infinitely many sets of positive rational numbers $x_j(1 \leq j \leq n)$ satisfying (1.8). Theorem 1.10 is proved.

From our proof, it is clear that the solution is constructible.

Example 1.5. When $n = 4$, we have the following positive rational solutions:

$$(1, 4, 2, 1), \left(1, \frac{49}{20}, \frac{128}{35}, \frac{25}{28}\right), \left(1, \frac{103058}{24497}, \frac{34969}{29737}, \frac{68644}{42449}\right)$$

from which the we obtain positive integer solutions

$$(778514660, 3114058640, 1557029320, 778514660),$$

$$(778514660, 1907360917, 2847139328, 695102375),$$

$$(778514660, 3275183240, 915488420, 1258930960).$$

They have the same sum 6228117280 and the same product $2938712953198523150291392472986880000$.

Example 1.6. When $n = 5$, we have the following positive rational solutions:

$$(1, 1, 1, 2, 5), \quad \left(1, 1, \frac{841}{221}, \frac{1690}{493}, \frac{289}{377}\right).$$

from which we obtain positive integer solutions

$$(6409, 6409, 6409, 12818, 32045), \quad (6409, 6409, 24389, 21970, 4913).$$

They have the same sum of 64090 and the same product 108131283474484110490.

Example 1.7. When $n = 6$, we have the following positive rational solutions:

$$(1, 1, 1, 2, 6), \quad \left(1, 1, \frac{1058}{273}, \frac{1323}{299}, \frac{388}{483}\right),$$

from which we obtain positive integer solutions

$$(6279, 6279, 6279, 6279, 12558, 37674).$$
$$(6279, 6279, 6279, 24334, 27783, 4394).$$

They have the same sum of 75348 and the same product 7354008786053535361179852.

It must be noted that Schinzel used Theorem 1.9 to prove the following

Theorem 1.11. *Given any positive integer k, there exist infinitely many primitive sets of k triples of positive integers with the same sum and the same product.*

Here primitive means that the members of the solution are relatively prime. We will use a similar approach and prove the following theorem using Theorem 1.10.

Theorem 1.12. *Given any positive integer k, there exist infinitely many sets of k n-tuples of positive integers such that they have the same sum and the same product.*

Proof of Theorem 1.9. Take any k positive rational number solutions of equation (1.7) $(x_{i1}, \ldots, x_{in}), i \leq k$, where $x_{i1} = \cdots = x_{i,n-3} = 1$. Denote

$$d = \text{lcm}_{i,j}(x_{i,j}, j = 1, 2, \ldots, n, i \leq k),$$

where lcm denotes the least common multiple. Let $x_{i,j} = \frac{a_{i,j}}{d}$, $a_{i,j}$ is a positive integer, $(gcd_{i,j}a_{i,j}, d) = 1$,

$$a_{i1} = \cdots = a_{i,n-3} = d$$

with

$$\sum_{i=1}^{n} a_{i,i} = 2nd, \quad \prod_{i=1}^{n} a_{i,i} = 2nd^n (i \leq k).$$

therefore

$$gcd_{i,j}a_{i,j} = 1.$$

Otherwise, let us have the prime $p|gcd_{i,j}, p \nmid d$, then it follows from (1.12) that $p^n|2n, n \geq 4$, impossible.

For the two sets of solutions $\{(x_{i1}, \ldots, x_{in}), i \leq k\}$ and $\{(x'_{i1}, \ldots, x'_{in}), i \leq k\}$, if the set of n-tuples $\{(a_{in}, \ldots, a_{in}), i \leq k\}$ and $\{(a'_{i1}, \ldots, a'_{in}), i \leq k\}$ coincide, then by (1.12) we have $d = d'$. As a result, the two sets of solutions also coincide. Therefore, by Theorem 1.10, for every positive integer k, there exist infinitely many sets of k n-tuples with the same sum and the same product. Theorem 1.12 is proved.

It is worth mentioning that using elliptic curve theory, we also proved the following two theorems, where the proof of Theorem 1.14 is similar to the proof of Theorem 1.12 (see Cai-Zhang, 2012).

Theorem 1.13. *When $n \geq 3$, the system of diophantine equations*

$$\begin{cases} \displaystyle\sum_{1 \leq l \leq m \leq n} x_l x_m = \dfrac{3n^2 - n - 2}{2} \\ x_1 \ldots x_n = 2n \end{cases}$$

admits infinitely many solutions in positive rational numbers.

The sum in Theorem 1.13 is known as a degree two elementary symmetric polynomial. In particular, when $n = 3$, the system of diophantine equations

$$\begin{cases} x_1 x_2 + x_2 x_3 + x_3 x_1 = 11 \\ x_1 x_2 x_3 = 6 \end{cases}$$

has infinitely many solutions in positive rational numbers. Three such solutions are

$$(3, 2, 1), \quad \left(\frac{91}{25}, \frac{65}{49}, \frac{610}{169}\right), \quad \left(\frac{487138}{152881}, \frac{437529}{426409}, \frac{255323}{139129}\right).$$

Theorem 1.14. *For every positive integer k, there exist infinitely many sets of k n-tuples positive integers such that they have the same product and take the same value under the degree two elementary symmetric polynomial.*

Remark 1.3. When $x_1, x_2, \ldots, x_{n-3}$ take any positive rational values, we can transform equation (1.8) into

$$\begin{cases} x_1 + x_2 + \cdots + x_n = a \\ x_1 x_2 \cdots x_n = b. \end{cases}$$

Here a, b are any given rational numbers, and a more general conclusion can be obtained in a similar way.

In 2012, I sent my first article on the new Waring problem to Professor Alan Baker (1940–2017) of Trinity College, Cambridge, an expert on the Waring problem and recipient of the 1974 Fields Medal. Soon after, I received an enthusiastic letter of encouragement from him, in which he praised it as a "truly original contribution" to the Waring problem. It was with the encouragement of Baker and others that we expanded the ideas of the additive and multiplicative equations to other problems in classical number theory and created in this way several new problems and research directions.

Chapter 2

The New Fermat Problem

I have a truly marvelous demonstration of this proposition, which this margin is too narrow to contain.

— *Pierre de Fermat*

2.1 Fermat's Last Theorem

In the previous chapter, we mentioned that the 18th century marginal note in Fermat's Latin edition of Diophantus's *Arithmetica* concerned the sum of four squares theorem, which states that every positive integer can be expressed as the sum of the squares of four integers. Fermat wrote a total of 48 marginal notes to Arithmetic, which were compiled and published by his son Samuel after his death (Fig. 2.1). The most famous of these is Fermat's Last Theorem, which was the second such note in order, and which he wrote in the margin of the eighth problem in Diophantus, concerning the solutions to the Pythagorean equation

$$x^2 + y^2 = z^2. \tag{2.1}$$

Positive integer solutions of (2.1) are called Pythagorean triples. As early as 3000 years ago, the Chinese knew that $(3, 4, 5)$ was the smallest Pythagorean triple. However, the Babylonians left clay tablets showing that they may have known about the existence of this triple and the Pythagorean theorem even earlier. When also $(x, y) = 1$, such solutions are called primitive Pythagorean triples.

Pierre de Fermat.

Figure 2.1.   Pierre de Fermat.

The book *Arithmetica* gives all primitive Pythagorean triples, which have the form

$$x = 2ab, \quad y = a^2 - b^2, \quad z = a^2 + b^2$$

where $a > b > 0$, $(a, b) = 1$, $a + b \equiv 1 \pmod 2$.

The proof of this result, which was given in Euclid's Elements and is therefore called Euclid's formula for Pythagorean triples, depends on the fact that if a square number is equal to the product of two relatively prime positive integers, then these two integers must themselves also be square numbers. With Euclid's formula, one obtains all positive integer solution of (2.1) by taking integer multiples of primitive solutions.

In particular, setting $a = n + 1, b = n$, we obtain an infinite number of solutions of (2.1) of the form

$$x = 2n + 1, \quad y = 2n^2 + 2n, \quad z = 2n^2 + 2n + 1.$$

According to Proclus (c. 410–485), the last major philosopher of ancient Greece, the above result is due to Pythagoras. He was particularly interested in the kind of triangles whose hypotenuse is longer than one of the right-angled sides by 1. For example, when n is taken to be 1,2 or 3, we obtain the solutions $(3, 4, 5), (5, 12, 13)$, and $(7, 24, 25)$, respectively.

The third smallest Pythagorean triple is $(8, 15, 17)$, which Proclus believed was discovered by Plato, who studied the integer triples of the form

$$x = 2n, \quad y = n^2 - 1, \quad z = n^2 + 1.$$

When $n > 1$ is even, this gives a Pythagorean triple by taking $a = n, b = 1$ in Euclid's formula. Such solutions correspond to triangles with hypotenuse longer by two than one of the right angled sides.

Fermat read through the solution of (2.1) and Euclid's formula carefully, and became interested in the case in which the exponent 2 of the equation is replaced with an exponent 3, 4, or 5. Even before then, Fermat had probably already considered the problem of determining Pythagorean triples and sums of squares from other perspectives. His seventh marginal note for example is rich in content, and can be divided into two parts.

The first part is about the representations of primes as sums of squares: if prime p is congruent to 1 modulo 4 (e.g., $5, 13, 17$, etc.), then there exist positive integers x, y such that

$$p = x^2 + y^2.$$

For example,

$$5 = 2^2 + 1^2, \quad 13 = 3^2 + 2^2, \quad 17 = 4^2 + 1^2.$$

On the other hand, if p is congruent to 3 modulo for $(3, 7, 11, \ldots)$ then the above equation does not even admit rational solutions.

The second part concerns the presentations of prime powers, and in particular squares of primes, as sums of squares.

In this case, if a prime p is congruent to 1 modulo 4, then there exists a right triangle with hypotenuse length p such that all three sides have integer lengths. If on the other hand, p is congruent to 3 modulo 4, then no such triangle exists.

For example,

$$5^2 = 4^2 + 3^2, \quad 13^2 = 12^2 + 5^2, \quad 17^2 = 15^2 + 8^2.$$

In the middle of the 20th century, this came to serve as a prelude to class field theory. Briefly stated, its solution exploits the relevant properties of the complex or imaginary numbers defined by $i = \sqrt{-1}$. Considering again the representation of primes as a sums of squares introduced by Fermat as above, we recognize:

$$5 = 2^2 + 1^2 = (2 + i)(2 - i),$$
$$13 = 3^2 + 2^2 = (3 + 2i)(3 - 2i),$$
$$17 = 4^2 + 1^2 = (4 + i)(4 - i).$$

Here, we are working in the ring $Z[i] = \{a + bi \mid a, b \in Z\}$ of Gaussian integers, in which $2 \pm i$, $3 \pm 2i$, $4 \pm i$ are "prime elements", playing a role equivalent to that of the prime numbers in the familiar ring Z of integers. According to the Fundamental Theorem of Arithmetic, every positive integer can be uniquely decomposed into the product of prime elements (up to the ordering of the factors). The fundamental theorem of arithmetic also holds in the ring of Gaussian integers (up to order or multiplication of the factors by the units $\pm 1, \pm i$). The primes congruent to 1 modulo decompose in this ring as a product of two conjugate prime elements, while the primes congruent to 3 modulo 4 remain prime $Z[i]$, and therefore cannot be decomposed into the product of two primes; these two facts express exactly the corresponding results about decomposition as a sum of squares above.

The problem of decomposing squares of primes as sums of squares similarly has a natural interpretation in terms of the arithmetic of $Z[i]$:

$$5^2 = (2 + i)^2(2 - i)^2 = (3 + 4i)(3 - 4i) = 3^2 + 4^2,$$
$$13^2 = (3 + 2i)^2(3 - 2i)^2 = (5 + 12i)(5 - 12i) = 5^2 + 12^2,$$
$$17^2 = (4 + i)^2(4 - i)^2 = (15 + 8i)(15 - 8i) = 8^2 + 15^2.$$

It was because of studies such as these on the Pythagorean Theorem that Fermat wrote his famous marginal note, around the year 1637:

> "It is impossible to separate a cube into two cubes, or a fourth power into two fourth powers, or in general, any power higher than the second, into two like powers."

On the side of this note, he penned an additional note:

> "I have discovered a truly marvelous proof of this, which this
> margin is too narrow to contain."

This is Fermat's Last Theorem, a problem that plagued generations of mathematicians until finally in 1995 the British mathematician Andrew Wiles (1953–) managed to prove it. Stating the problem
symbolically, the equation

$$x^n + y^n = z^n, \; n \geq 3 \tag{2.2}$$

has no solutions in positive integers.

It is obvious that we only need only consider the case where n
is an odd prime. As for the idea of Wiles' proof, we describe it in
Section 2.4. It is worth mentioning that after the formulation of Fermat's Last Theorem, Fermat did not turn to the study of higher
degree equations, but remained focused on quadratic equations. For
example, in a letter to Pascal on September 24, 1654, Fermat, who
was already in his later years, stated that

Each prime number of the form $3k + 1$ can be written as $x^2 +
3y^2 x^2 + 3y^2$ and every prime number of the form $8k + 1$ or $8k + 3$
can be written as $x^2 + 2y^2 x^2 + 2y^2$.

For example,

$$7 = 2^2 + 3 \cdot 1^2, \; 13 = 1^2 + 3 \cdot 2^2, \quad 19 = 4^2 + 3 \cdot 1^2$$

and

$$11 = 3^2 + 2 \cdot 1^2, \quad 17 = 3^2 + 2 \cdot 2^2, \quad 19 = 1^2 + 2 \cdot 3^2.$$

Regarding the proof of his theorem to which Fermat alluded in his
note, naturally no one has ever seen it. And judging from the tools
and difficulties involved in the eventual proof given by Wiles, any
proof seems far beyond the mathematics of the time. Fermat did give
a proof of the special case $n = 4$, making use of his method of infinite
descent and twice invoking Euclid's formula; his proof appears in his
Arithmetica marginalia.

In fact, Fermat proves a stronger result, namely that the equation

$$x^4 + y^4 = z^2, \quad (x, y) = 1$$

has no solutions in positive integers. For suppose there exists some solution; then it follows from Euclid's formula that

$$x^2 = 2st, \quad y^2 = s^2 - t^2, \quad z = s^2 + t^2, \quad s > t, \quad (s, t) = 1.$$

We have

$$t^2 + y^2 = s^2, \quad (t, y) = 1$$

where y and s are odd, t is even, and $(s, 2t) = 1$.

From $x^2 = s(2t)$, we have $s = u^2, 2t = v^2$. Then using Euclid's formula again,

$$t = 2ST, \quad y = S^2 - T^2, \quad s = S^2 + T^2, \quad (S, T) = 1$$

Therefore, there are

$$\left(\frac{v}{2}\right)^2 = ST, \quad S = X^2, \quad T = Y^2$$

thus

$$X^4 + Y^4 = S^2 + T^2 = s = u^2,$$
$$(X, Y) = 1.$$

Let $Z = u$. Then $Z = u \leq u^2 = s < s^2 + t^2 = z$. This is a contradiction insofar as it implies an infinite descent of positive integers.

We can also prove directly that (2.2) has no positive integer solutions at $n = 4$ via a different application of infinite descent, for which we only need to prove that the elliptic curve

$$y^2 = x^3 - x \qquad (2.3)$$

has only three rational number points, given by $(0, 0)$ and $(\pm 1, 0)$.

This is because if (2.2) has a positive integer solution for $n = 4$, multiplying y^4, shuffling terms and multiplying both sides by $\frac{z^2}{y^6}$ gives

$$\left(\frac{x^2 z}{y^8}\right)^2 = \left(\frac{z^2}{y^2}\right)^3 - \frac{z^2}{y^2}.$$

This implies that there exist rational pairs (x, y) with $y \neq 0$ that satisfy equation (2.3).

In the following, we briefly sketch the idea of the proof.

For a rational number $a = \frac{m}{n}, gcd(m, n) = 1$, define its height as $H(a) = \max(|n|, |m|)$. If (2.3) has rational points other than $(0, 0)$ and $(\pm 1, 0)$ other than rational number solutions, choose the one with the smallest x-coordinate height, which we denote (x_0, y_0). It is possible to show that this implies the existence of rational points other than $(0, 0)$ and $(\pm 1, 0)$ with x-coordinate height smaller than x_0, contradiction.

In 1753, Euler, a guest in Berlin, wrote to Goldbach, who was working at the Ministry of Foreign Affairs in Moscow, that he proved Fermat's conjecture for $n = 3$, but his proof did not appear in his own book, *Elements of Algebra*, until 1770, after his return to St. Petersburg. The proof makes use again of Fermat's method of infinite descent, working in the ring of integers of the imaginary quadratic field $Q(\sqrt{-3})$ and taking advantage of congruence results and the unique factorization property (see Hardy–Wright, 1979, 13.4). In fact, Euler proved a much stronger result, namely that the following equation:

$$\alpha^3 + \beta^3 + \gamma^3 = 0$$

has no non-trivial solutions in the ring of integers of $Q(\sqrt{-3})$, i.e., no solutions satisfying $\alpha \neq 0, \beta \neq 0, \gamma \neq 0$. We give a concise proof of Fermat's conjecture for $n = 3$ using the method of elliptic curves in Chapter 4, where we discuss variants of Euler's conjecture.

In 1816, the Paris Academy of Sciences decided that Fermat's proposed conjecture should almost certainly prove to be valid, and named it Fermat's Last Theorem (to distinguish it from Fermat's Little Theorem); they set up a prize and medal for the anyone who could prove it, and from that point on Fermat's Last Theorem became famous to the world.

Figure 2.2. Sophie Germain Building(right) in Université Paris Cité, photograph by the author.

The French mathematician Sophie Germain (1776–1831) proved that when n and $2n + 1$ are prime numbers (later called Sophie Germain prime numbers), at least one of x, y, and z satisfying Fermat's equation is a multiple of n (Fig. 2.2). On this basis, in 1825, the 20-year-old German mathematician P. G. L. Dirichlet (1805–1859) and the French mathematician A.-M. Legendre (1752–1833) independently proved that Fermat's Last Theorem holds in the case $n = 5$, using an extension of Euler's method, but avoiding the use of unique factorization. In 1839, the French mathematician Gabriel

Lamé (1795–1870) further improved Germain's method, proving the case of $n = 7$.

In 1844, the German mathematician Ernst Kummer (1810–1893) introduced the concept of "ideal numbers", which led to an important breakthrough in the study of Fermat's Last Theorem. Three years later, he introduced the concept of regular prime numbers; in modern terminology, a prime p is regular if it does not divide the class number of the cyclotomic field $Q(\zeta_p)$ of the class number of. Kummer proved that (2.2) has no solution for regular primes. He also verified that among the prime numbers less than 100, only 37, 59, and 67 are irregular.

In 1847, the Paris Academy of Sciences staged a dramatic scene. Mathematicians Lamé and Cauchy successively announced the proof of Fermat's Last Theorem. Lamé claimed that the proof used the unique factorization property of cyclotomic fields, which he attributed to Joseph Liouville (1809–1882). Liouville said in turn that this theorem originated from the ideas of Euler and Gauss. The conclusion seemed to be very reliable, but at this time, Liouville received a letter from Kummer, pointing out that the unique factorization property of cyclotomic fields does not in fact hold in general.

Later in the 20th century, the French mathematician Henri L. Lebesgue (1875–1941) also submitted a proof of Fermat's Last Theorem to the Paris Academy of Sciences, which also turned out to be wrong.

In 1908, the Royal Scientific Society of Göttingen announced the Wolfskehl Prize: 100,000 marks would be awarded to anyone who solved Fermat's Last Theorem within 100 years. Wolfskehl was a German industrialist, who had decided to commit suicide when he was young. On the eve of his intended suicide, he read Kummer's exposition of the errors in the proof of the Fermat theorem by Lamé and Cauchy, and could not help calculating until dawn and gave up his obsession; and mathematics brought about a rebirth in his spirits, and later he became a great tycoon.

Meanwhile, after tedious case by case calculations, Kummer and one of his colleagues verified that (2.2) has no solutions for the irregular prime numbers 37, 59, and 67. This analysis unfortunately becomes more and more difficult to carry out for larger irregular primes. Twenty-two years after Kummer's death, it was found that there are infinitely many irregular prime numbers, and it remains unknown today whether or not there are infinitely many regular prime numbers.

Fortunately, with the availability of electronic computers and methods like the Lehmer congruence equation, by the 1980s, Samuel S. Wagstaff of the University of Illinois had extended the range of completed cases to all to $n \leq 25,000$, and by the time Wiles proved Fermat's Last Theorem, it had been verified that equation (2.2) has no solutions for all n up to 4 million. But in any case, even the most powerful computers can only verify a finite amount of data and certainly cannot prove that equation (2.2) is unsolvable for all prime exponents p.

Wiles' proof came at the right time, and he claimed the prize left by Wolfskehl. It was an epoch-making achievement, the most important advance in the field of number theory in the 20th century, and has even been called the "mathematical achievement of the 20th century", just as the proof of the prime number theorem in 1896 was called the "mathematical achievement of the 19th century". They are considered the "great white sharks" of mathematics, and for this reason Wiles was awarded a special IMU Silver Plaque in lieu of the Fields Medal, which he was already too old at that time to receive.

In fact, what Wiles proved is called the Taniyama–Shimura conjecture, proposed in 1955 by two Japanese mathematicians, Yutaka Taniyama (1927–1958) and Goro Shimura (1930–2019) (Fig. 2.3),

Figure 2.3.   Office of Shimura, photograph by the author in Princeton.

which states that all elliptic curves over the domain of rational numbers are moduli curves. In 1986, the German mathematician Gerhard Frey had proposed the so-called "ε conjecture": if Fermat's Last Theorem does not hold, then elliptic curves

$$y^2 = x(x - a^n)(x + b^n)$$

would be a counterexample to the Taniyama–Shimura conjecture. Frey's conjecture was confirmed a little later by the American mathematician Kenneth Ribet, thus revealing the close connection between Fermat's Last Theorem and elliptic curves and modular forms, and verifying that the Taniyama–Shimura conjecture has Fermat's Last Theorem as a corollary.

Several famous extensions of Fermat's Last Theorem have been made, such as Beal's Conjecture. Andrew Beal (1952–), an American businessman and banker obsessed with number theory, was the first to conjecture in 1993 that when a, b, and c are all integers greater than or equal to 3, the equation

$$x^a + y^b = z^c, (x, y, z) = 1$$

has no positive integer solutions.

The coprimality condition of the above equation is necessary, since otherwise there are many counterexamples, such as $3^3 + 6^3 = 3^5, 7^6 + 7^7 = 98^3$. In addition, there are general formulas

$$[a(a^m + b^m)]^m + [b(a^m + b^m)]^m = (a^m + b^m)^{m+1}$$

where a and b are arbitrary positive integers, $m \geq 3$. In particular, taking $a = b = 1$, we have $2^m + 2^m = 2^{m+1}$.

Another example is that in 1995, the French-Canadian mathematician Darman and the British mathematician Granville (cf. Darman-Granville, 1995) came up with the following:

Fermat–Catalan Conjecture. Only finitely many mutually exclusive groups of three numbers $\{x^p, y^q, z^r\}$ satisfy the equation

$$x^p + y^q = z^r \quad (\mathrm{F} - \mathrm{C})$$

where p, q, r are positive integers satisfying $\frac{1}{p} + \frac{1}{q} + \frac{1}{r} < 1$.

So far, only ten solutions have been found, namely

$$1 + 2^3 = 3^2, \quad 2^5 + 7^2 = 3^4, \quad 13^2 + 7^3 = 2^9,$$

$$2^7 + 17^3 = 71^2, \quad 3^5 + 11^4 = 122^2, \quad 33^8 + 1549034^2 = 15613^3,$$

$$1414^3 + 2213459^2 = 65^7, \quad 9262^3 + 15312283^2 = 113^7,$$

$$17^7 + 76271^3 = 21063928^2, \quad 43^8 + 96222^3 = 30042907^2.$$

It is noteworthy that each solution contains exactly one squared term. In other words, no set of solutions has been found whose every exponent is greater than or equal to 3. In other words, they are not counterexamples to Beal's conjecture.

Obviously, the Beal equation is a special case of the Fermat–Catalan equation. If the abc conjecture holds, it can be shown that there are at most finitely many counterexamples to either Beal's conjecture or the Fermat–Catalan conjecture. It is worth mentioning that in 2013, Beal, himself set up a prize of \$1 million for the first person to prove or disprove his conjecture.

2.2 The New Fermat Problem

In October 2011, inspired by their work on the New Waring problem, the author formulated the New Fermat problem, which considers positive integer solutions of the equations

$$\begin{cases} A + B = C \\ ABC = x^n. \end{cases} \tag{2.4}$$

When $d = gcd(A, B, C) = 1$, equation (2.4) is equivalent to Fermat's Last Theorem. That is, (2.4) has no positive integer solution.

Let $\omega(n)$ denote the number of distinct prime factors of n (when $d = 1, \omega(d) = 0$). We consider the case of $\omega(d) = 1$ and $n > 3$ is a prime number, say p. The data show that $d = q^\alpha$, where α satisfies $3\alpha + 1 \equiv 0 \pmod{p}$. Taking $p = 5$ as an example, we investigate the form of q.

Let a be a positive integer, b be an integer, with $(a, b) = 1$ and $a + b = m^5$ such that $\frac{a^5 + b^5}{a+b} = q$ is prime. Note that

$$q = (a^3 - b^3)(a - b) + a^2 b^2 \geq \max\{a^2, b^2\}.$$

Therefore $(q, a) = (q, b) = 1$, so that if

$$A = q^3 a^5,$$
$$B = q^3 b^5,$$
$$C = q^4 (a + b).$$

then (A, B, C) satisfies (2.1) and $d = (A, B, C) = q^3$.

For example, taking $a = n + 1, b = -n$, then $q = (n + 1)^5 - n^5$, giving $q = 31, 211, 4651, 61051, 371281, \ldots$. On the other hand, with $a_1 + b_1 = c_1^{5r}$, say r = 1, and $c_1 = 2$, then we have the solution

$$\{a_1, b_1, q, c_1\} = \{11, 21, 132661, 2\}, \{9, 23, 202981, 2\}.$$

Note than 31 and 132661 are the smallest generalized Mersenne prime and non-generalized Mersenne prime, respectively, such that (2.4) has a solution and $d = 31^\alpha, 132661^\alpha$, where $3\alpha + 1 \equiv 0 \pmod 5$.

We (Cai *et al.*, 2015, which is referenced in the English Wikipedia entry "Fermat's Last Theorem") have carried out some research into the New Fermat problem.

Theorem 2.1. *Let $n \geq 3$, if $n \equiv 0 \pmod 3$, then (2.4) has no positive integer solutions; if $n \not\equiv 0 \pmod 3$, then (2.4) has infinitely many positive integer solutions.*

Proof. If $n \not\equiv 0 \pmod 3$, then there exists a positive integer k such that $3k + 2 \equiv 0 \pmod n$. By the properties of Pythagorean triples, there exists an infinite number of positive integers (a, b, c) satisfying $a^2 + b^2 = c^2$. Let

$$\begin{cases} A = a^{k+2} b^k c^k, \\ B = a^k b^{k+2} c^k, \\ C = a^k b^k c^{k+2}. \end{cases}$$

Then (A, B, C) satisfies (2.4), where $D = (abc)^{\frac{3k+2}{n}}$.

Also if $n \equiv 0 \pmod 3$, suppose there is a triplet (A, B, C) satisfying (2.4). Let $d = gcd(A, B, C)$. Then $\gcd\left(\frac{A}{d}, \frac{B}{d}, \frac{C}{d}\right) = 1, d^3 \mid D^n$.

But $3 \mid n$, and so $d \mid D^{\frac{n}{3}}$, and

$$\begin{cases} \frac{A}{d} + \frac{B}{d} = \frac{C}{d} \\ \frac{A}{d} \cdot \frac{B}{d} \cdot \frac{C}{d} = \left(\frac{D^{n/3}}{d} \right)^3. \end{cases}$$

Thus, we have $\frac{A}{d} = x^3, \frac{B}{d} = y^3, \frac{C}{d} = z^3, x^3 + y^3 = z^3$, which contradicts Fermat's Last Theorem. Theorem 2.1 is proved. □

Theorem 2.2. *Let* $\gcd(A, B, C) = p^k, k \geq 1$. *When* $n = 4$, *if* p *is an odd prime*, $p \equiv 3 \pmod{8}$, *then* (2.4) *has no positive integer solutions; when* $n = 5$, *if* $p \not\equiv 1 \pmod{10}$, *then* (2.4) *has no positive integer solutions.*

Remark 2.1. *For* $p = 2$, *or* $p \equiv 1, 5, 7 \pmod{8}$, (2.4) *may have positive integer solutions satisfying* $\gcd{pB}, C$. *For example.*

$$\begin{cases} 2 + 2 = 4 \\ 2 \times 2 \times 4 = 2^4, \end{cases} \quad \begin{cases} 17 + 272 = 289 \\ 7 \times 272 \times 289 = 34^4, \end{cases} \quad \begin{cases} 5 + 400 = 405 \\ 5 \times 400 \times 405 = 30^4, \end{cases}$$

$$\begin{cases} 47927607119 + 1631432881 = 49559040000 \\ 47927607119 \times 1631432881 \times 49559040000 = 4436760^4. \end{cases}$$

This is where $\gcd(A, B, C)$ is equal to the prime numbers $2, 17, 5$, and 239 respectively.

To prove Theorem 2.2, we need two lemmas:

Lemma 2.1. *Let* p *be a prime and* $n \geq 2$ *an integer. If* $\gcd(A, B, C) = p^k, k \geq 1$, *and* $k \equiv 0 \pmod{n}$, *then* (2.4) *has no nonzero integer solutionss.*

Proof. Let $A_1 = \frac{A}{p^k}, B_1 = \frac{B}{p^k}, C_1 = \frac{C}{p^k}$. Then (2.4) becomes

$$\begin{cases} A_1 + B_1 = C_1 \\ p^{3k} A_1 B_1 C_1 = D^n. \end{cases}$$

But, $k \equiv 0 \pmod{n} A_1, B_1, C_1$ are relatively prime, so $A_1 = x^n, B_1 = y^n, C_1 = z^n, x^n + y^n = z^n$. By Fermat's Last Theorem, we must have, so $xyz = 0 ABC = 0$, contradiction. Lemma 2.1 is proved. □

Lemma 2.2. *If $A > 2$ is a positive integer and A has no prime factor of the form $10k + 1$, then*

$$\begin{cases} x^5 + y^5 = Az^5 \\ gcd(x, y) = 1 \end{cases}$$

has are no non-zero integer solutions. If $A = 2$, then the solution to the above equation is $(x, y, z) = \pm(1, 1, 1)$.

Lemma 2.2 *was formulated as a conjecture in 1843 by the French mathematician V. A. Lebesgue (not to be confused with the founder of modern real analysis, H. L. Lebesgue). It was only proved in 2004, by Halberstadt and Kraus (2004).*

Proof of Theorem 2.2. We start with the first claim of the theorem. By Lemma 2.1, we only need to discuss the case $k \not\equiv 0 (\mathrm{mod}\ 4)$. Consider an odd prime $p \equiv 3 (\mathrm{mod}\ 8)$, such that (2.4) has solution (A, B, C). Considering $gcd(A, B, C) = p^k$, (2.4) becomes

$$\begin{cases} A_1 + B_1 = C_1, \\ p^{3k} A_1 B_1 C_1 = D^4, \end{cases} \tag{2.5}$$

where $A_1 = \frac{A}{p^k}, B_1 = \frac{B}{p^k}, C_1 = \frac{C}{p^k}$ are relatively prime. Since $k \not\equiv 0 (\mathrm{mod}\ 4)$, one and only one of A_1, B_1, and C_1 is divisible by p. Let $3k \equiv r (\mathrm{mod}\ 4), 1 \leq r \leq 3$. By (2.5) (and interchanging A_1 with B_1 if necessary), we have

$$\begin{cases} A_1 = x^4, \\ B_1 = y^4, \\ C_1 = p^{4-r} z^4, \end{cases} \tag{2.6}$$

or

$$\begin{cases} A_1 = p^{4-r} z^4, \\ B_1 = y^4, \\ C_1 = x^4, \end{cases} \tag{2.7}$$

where x, y and pz are relatively prime in pairs.

If A_1, B_1 and C_1 satisfy (2.6), then

$$x^4 + y^4 = p^{4-r} z^4. \tag{2.8}$$

Since $gcd(y, p) = 1$, there exists $s \not\equiv 0 \pmod{p}$, such that $sy \equiv 1 \pmod{p}$. From (2.8), we therefore obtain

$$(xs)^4 \equiv -1 \pmod{p}.$$

This implies that -1 is a quadratic residue modulo p, which implies that either $p = 2$ or $p \equiv 1 \pmod 4$, contradiction.

Next, if A_1, B_1 and C_1 satisfy (2.7), then

$$x^4 - y^4 = p^{4-r} z^4. \tag{2.9}$$

If r $= 2$, then we have $x^4 - y^4 = (pz^2)^2$. On the other hand, it has long been known that the equation $X^4 - Y^4 = Z^2$ has no non-zero integer solutions. Therefore, setting r $= 1$ or 3, (2.9) becomes in turn

$$x^4 - y^4 = p(pz^2)^2$$

or

$$x^4 - y^4 = p \left(z^2\right)^2.$$

Now we make use of two properties of congruent numbers, the definition of which is given in Chapter 5 . First, a prime p is not a congruent number when $p \equiv 3 \pmod 8$. and second, if the equation $x^4 - y^4 = cz^2$ has a solution with $xyz \neq 0$, then $|c|$ must be a congruent number.

With these facts in hand, we turn to the second claim of the theorem. By Lemma 2.1, we only need to discuss the case of $k \not\equiv 0 \pmod 5$. Noting that $gcd(A, B, C) = p^k$, (2.4) can be transformed into

$$\begin{cases} A_1 + B_1 = C_1, \\ p^{3k} A_1 B_1 C_1 = D^5, \end{cases} \tag{2.10}$$

where $A_1 = \frac{A}{p^k}, B_1 = \frac{B}{p^k}, C_1 = \frac{C}{p^k}$ are pairwise relatively prime. Since $k \not\equiv 0 \pmod 5$, one and only one of A_1, B_1 and C_1 is divisible by p. Let $3k \equiv r \pmod 5, 1 \leq r \leq 4$. From (2.10), (and permuting A_1, B_1, and C_1 if necessary), we have

$$\begin{cases} A_1 = x^5, \\ B_1 = y^5, \\ C_1 = p^{5-r} z^5, \end{cases} \tag{2.11}$$

where x, y, and pz are pairwise relatively prime. From (2.10) and (2.11), we get

$$x^5 + y^5 = p^{5-r} z^5. \tag{2.12}$$

However, the odd prime number $p \not\equiv 1 \pmod{10}$, by Lemma 2.1, (2.12) has no positive integer solutions. Theorem 2.2 is proved.

Following upon this result, we also formulate the following conjectures for p be an odd prime.

Conjecture 2.1. *If $\gcd(A, B, C) = p$ and n is an odd prime number, then equation (2.4) has no solutions in nonnegative integers.*

Conjecture 2.2. *If $\gcd(A, B, C) = p^k, k \geq 1$, and n is an odd prime numbers, then (2.4) has no solutions in non-negative integers when $p \not\equiv 1 \pmod{2n}$.*

Conjecture 2.3. *If $n > 3$ is a prime number, $n \equiv r \pmod{3}$, $1 \leq r \leq 2$, and $\gcd(A, B, C) = p^k$, where k is a positive integer, p is an odd prime, and $p \not\equiv 1 \pmod{2n}$, then (2.4) has no solutions in positive integers.*

For $n = 3$, we can confirm that Conjecture 2.1 is correct by Theorem 2.1.

Remark 2.2. If the abc conjecture holds, then Conjecture 2.1 is true for fixed p and sufficiently large n. Here n need not be a prime number.

Proof. If $\gcd(A, B, C) = p$, then (2.4) becomes

$$\begin{cases} \frac{A}{p} + \frac{B}{p} = \frac{C}{p}, \\ ABC = D^n, \end{cases}$$

so that

$$\mathrm{rad}\left(\frac{ABC}{p^3}\right) = \mathrm{rad}\left(\frac{D^n}{p^3}\right) = \mathrm{rad}\left(\left(\frac{D}{p}\right)^3 D^{n-3}\right)$$

$$\leq \mathrm{rad}(D), C > D^{n/3}.$$

For an arbitrary $n \geq 7, 0 < \varepsilon < \frac{1}{3}$, we have

$$p \leq D = D^{\frac{7}{3}-1-\frac{1}{3}} \leq D^{\frac{7}{3}-1-\varepsilon}$$

Therefore,

$$q\left(-\frac{A}{p}, -\frac{B}{p}, \frac{C}{p}\right)$$

$$= \frac{\log\left(\frac{C}{\rho}\right)}{\log\left(\text{rad}\left(\frac{ABC}{p^3}\right)\right)}$$

$$\geq \frac{\frac{n}{3}\log D - \left(\frac{3}{7} - 1 - \varepsilon\right)\log D}{\log D}$$

$$\geq 1 + \varepsilon.$$

By the third form of the *abc* conjecture, there exists only finitely many triplets $\left(-\frac{A}{p}, -\frac{B}{p}, \frac{C}{p}\right)$. Let $A_1 = \frac{A}{p}, B_1 = \frac{B}{p}, C_1 = \frac{C}{p}$, then $gcd(A_1, B_1, C_1) = 1$, $p^3 A_1 B_1 C_1 = D^n$. Let M be the largest such integer m. There exists a prime q such that $q^m \mid A_1 B_1 C_1$. Thus, if $n > M + 3$, then $p^3 A_1 B_1 C_1 = D^n$ has no solution. It follows when $gcd(A, B, C) = p$, (2.4) has no solutions for sufficiently large positive integers n. □

Finally, if $n > 3$ is prime, let us construct the prime p such that (2.4) has positive integer solutions (A, B, C) satisfying $gcd(A, B, C) = p^k$.

Theorem 2.3. *If $n > 3$ is a prime number, $n \equiv r(\text{mod } 3), 1 \leq r \leq 2, a, b, m \neq 0$ are integers such that $\frac{a^n + b^n}{a + b} = p$ is an odd prime, $a + b = m^n$, then for $p \equiv l(\text{mod } 2n)$, equation (2.4) has positive integer solutions satisfying $gcd(A, B, C) = p^k$, where the positive integer $k \equiv \frac{rn-1}{3}(\text{mod } n)$.*

In particular, let $a = 2, b = -1, m = 1$, we have

Corollary 2.1. *If $n > 3$ is a prime number, $n \equiv r(\text{mod } 3), 1 \leq r \leq 2$, and $p = 2^n - 1$ is a Mersenne prime, then there exists a positive integer solution of (2.4) satisfying $gcd(A, B, C) = p^k$, where the positive integer $k \equiv \frac{rn-1}{3}(\text{mod } n)$.*

Proof of Theorem 2.3. First, we prove that $f(x, y) = \frac{x^k + y^k}{x + y} > 0$, where k is a positive odd number, holds for any real numbers x and y, satisfying $x + y \neq 0$. Clearly, $f(x, y) > 0$ when $xy \geq 0$. Assume below that $xy < 0$ and use induction for odd k. When $k = 1$, $f(x, y) = 1 > 0$. Assume that holds for positive odd k, where $x + y \neq 0$. We have

$$\frac{x^{k+2} + y^{k+2}}{x + y} = \frac{x^k + y^k}{x + y}(-xy) + x^{k+1} + y^{k+1} > 0.$$

By the inductive hypothesis, the conclusion holds.

By assumption, $n > 3$, so

$$0 < p = \frac{a^n + b^n}{a + b} = a^{n-1} - a^{n-2}b + a^{n-3}b^2 - \cdots + b^{n-1},$$

Therefore, $gcd(p, a) = gcd(p, b) = gcd(a, b) = 1$ and $gcd(a, a+b) = gcd(b, a + b) = 1$. Set

$$\begin{cases} A = p^{\frac{rn-1}{3} + tn} a^n \\ B = p^{\frac{rn-1}{3} + tn} b^n \\ C = p^{\frac{rn+2}{3} + tn}(a + b). \end{cases}$$

Here r is as defined in the statement of the theorem and t is a non-negative integer.

Noting that $a + b = m^n$, it follows that A, B, C satisfy (2.4), where $x = p^{r+3t} abm$,

$$gcd(A, B, C) = p^k,$$

where the positive integer $k \equiv \frac{rn-1}{3} + tn \equiv \frac{rn-1}{3} (\text{mod } n)$.

Finally, let us prove that $p \equiv 1 (\text{mod } 2n)$. Since both p and n are odd prime numbers, we only need to prove that

$$p \equiv 1 (\text{mod } n).$$

From $\frac{a^n + b^n}{a + b} = p$, n is an odd prime, and from Fermat's little theorem it follows that

$$p(a + b) = a^n + b^n \equiv a + b (\text{mod } n).$$

Thus, it is sufficient to prove that $(p, a + b) = 1$, and conversely let $a + b \equiv 0 (\text{mod } n)$, then $-b \equiv a (\text{mod } n)$.

$p = \frac{a^n + b^n}{a + b} = a^{n-1} - a^{n-2}b + a^{n-3}b^2 - \cdots + b^{n-1} \equiv na^{n-1} \equiv 0$ (mod n). Since both p and n are prime numbers, we get $p = n$.

Table 2.1. $n = 5$, prime numbers p satisfying Theorem 2.3, $p < 10^7$.

p	a	b	m				
				1641301	35	-3	2
31	2	-1	1	1803001	25	-24	1
211	3	-2	1	2861461	28	-27	1
4651	6	-5	1	4329151	31	-30	1
61051	11	-10	1	4925281	32	-31	1
132661	11	21	2	5754901	45	-13	2
202981	9	23	2	7086451	35	-34	1
371281	17	-16	1	7944301	36	-35	1
723901	20	-19	1	8782981	49	-17	2

The following is divided into two cases. If $ab < 0$, it may be useful to set $a > 0$ and $b < 0$, then

$p = \frac{a^n + b^n}{a+b} = a^{n-1} - a^{n-2}b + a^{n-3}b^2 - \cdots + b^{n-1} > n = p$, contradiction.

If $ab \geq 0$, it may be useful to set $a \geq 0$ and $b \geq 0$. Noting that $p = \frac{a^n + b^n}{a+b}$, we have $a \neq b$ and $ab \neq 0$. By symmetry, we can set $a \geq 1$ and $b \geq 2$.

When $a = 1, b \geq 2$, the $p = \frac{1+b^n}{1+b} = \frac{1+b^p}{1+b}$. Let $f(b) = b^p + 1 - p(b+1)$, then we have

$$f'(b) = pb^{p-1} - p = p(b^{p-1} - 1) > 0.$$

Therefore,

$$f(b) \geq f(2) = 2^p + 1 - 3p > 0,$$

where p is an odd prime. Therefore $\frac{1+b^p}{1+b}\frac{1+b^p}{1+b} > p$, contradiction.

When $a \geq 2, b \geq 2, p = \frac{a^n + b^n}{a+b} = \frac{a^p + b^p}{a+b} > \frac{pa+pb}{a+b} = p$, contradiction.

Theorem 2.3 is proved.

In the following, for $n = 5, 7$, we list some solutions for primes p that satisfy the conditions of Theorem 2.3.

It is worth noting that if a and b are one positive and one negative, the negative term can be moved to the other side of the equation and still be a set of positive integer solutions to (2.4).

Table 2.2. $n = 7$, prime numbers p satisfying Theorem 2.3, $p < 10^{11}$.

p	a	b	m	p	a	b	m
127	2	-1	1	1928294551	26	-25	1
14197	4	-3	1	8258704609	33	-32	1
543607	7	-6	1	14024867221	36	-35	1
1273609	8	-7	1	22815424087	39	-38	1
2685817	9	-8	1	30914273881	41	-40	1
5217031	10	-9	1	77617224511	59	69	2
16344637	12	-11	1	91154730577	49	-48	1
141903217	17	-16	1	98201826199	55	73	2

2.3   Other Number Fields

Since Fermat's equation has no non-zero solutions in the field of rational numbers, the search for its solutions in other fields was started, with the richest results in the study of its solutions in quadratic fields. As early as 1915, W. Burnside found the solution of Fermat's equation $(n = 3)$

$$\begin{cases} x = -3 + \sqrt{-3\left(1 + 4k^3\right)} \\ y = -3 - \sqrt{-3\left(1 + 4k^3\right)} \\ z = 6k, \end{cases}$$

where k is a rational number not equal to $0, -1$. When $k = 0$, the Fermat equation has no non-zero solutions in the quadratic field $Q(\sqrt{-3})$. Since then, there has been a lot of research in this direction.

Almost a century later, in 2013, M. Jones and J. Rouse gave, under the assumption of the BSD conjecture, a sufficient condition for the Fermat equation to have a non-zero solution in the quadratic field $Q(\sqrt{t})$, where t is a squarefree integer.

Considering equation (2.4) for $n = 3$ in the quadratic field $Q(\sqrt{t})$, we obtain the following non-zero solution.

Proposition. *If $t \neq 0, -1$ is a squarefree integer, such that the elliptic curve*

$$tu^2 = 1 + 4k^3$$

The Fermat equation on French stamp

The Fermat monument at his hometown of Toulouse

has non-zero rational points (u, k), *then when* $n = 3$, (2.4) *has infinitely many non-zero solutions* (A, B, C, x) *in the quadratic field* $Q(\sqrt{t})$.

Proof. Let $A = a + b\sqrt{t}$ $B = c + d\sqrt{t}$ $x = e + f\sqrt{t}$ When $n = 3$ is taken in (2.4) we get

$$a^2c + ac^2 + 2adtb + ad^2t + b^2tc + 2btcd$$
$$+ (2acb + 2acd + a^2d + bc^2 + tb^2d + tbd^2)$$
$$\sqrt{t} = e^3 + 3ef^2t + f(3e^2 + f^2t)\sqrt{t}$$

therefore

$$\begin{cases} a^2c + ac^2 + 2adtb + ad^2t + b^2tc + 2btcd = e^3 + 3ef^2t, \\ 2acb + 2acd + a^2d + bc^2 + tb^2d + tbd^2 = f(3e^2 + f^2t). \end{cases}$$

Solving the above two equations yields

$$t = -\frac{a^2c + ac^2 - e^3}{ad^2 + b^2c + 2bcd + 2adb - 3ef^2}$$
$$= -\frac{2acb + 2acd + a^2d + bc^2 - 3fe^2}{b^2d + bd^2 - f^3}.$$

Taking $e = kc$, $f = kd$, from the expression of t we get the equation

$$(ad + 2ab + cb - 2ck^3d)(d^2a^2 + c^2b^2 + 2cbda$$
$$+ 2c^2db + 2cd^2a - 4c^2d^2k^3) = 0.$$

Considering $ad + 2ab + cb - 2ck^3d = 0$, the solution is

$$d = -\frac{b(c + 2a)}{a - 2ck^3}.$$

Therefore.

$$t = \frac{(a - 2ck^3)^2}{(1 + 4k^3)b^2}.$$

Let $a - 2ck^3 = (1 + 4k^3)b$, we have

$$t = 1 + 4k^3.$$

This gives the solution

$$\begin{cases} A = \left(1 + 4k^3\right)a + \left(a - 2k^3c\right)\sqrt{1 + 4k^3} \\ B = \left(1 + 4k^3\right)c - (2a + c)\sqrt{1 + 4k^3} \\ C = \left(1 + 4k^3\right)(a + c) - \left(a + \left(2k^3 + 1\right)c\right)\sqrt{1 + 4k^3} \\ x = \left(1 + 4k^3\right)kc - k(2a + c)\sqrt{1 + 4k^3}, \end{cases}$$

where $a, c, k \in Q - \{0\} \in Q - \{0\}$.

Let $tu^2 = 1 + 4k^3$, where $t\backslash = 0, -1$ is a squarefree integer. Then for $n = 3$, we get the non-zero solution

$$\begin{cases} A = uta + \left(a - 2k^3c\right)\sqrt{t} \\ B = utc - (2a + c)\sqrt{t} \\ C = ut(a + c) - \left(a + \left(2k^3 + 1\right)c\right)\sqrt{t} \\ x = utkc - k(2a + c)\sqrt{t}. \end{cases}$$

$\square$

Example 2.1. When $k = -1, t = -3$, Therefore, (2.4) has solutions for $n = 3$ in the quadratic field $Q\left(\sqrt{-3}\right)$ given by

$$\begin{cases} A = -3ua + (a + 2c)\sqrt{-3} \\ B = -3uc - (2a + c)\sqrt{-3} \\ C = -3u(a + c) - (a - c)\sqrt{-3} \\ x = 3uc + (2a + c)\sqrt{-3}. \end{cases}$$

To obtain a Burnside-like solution for any given t, we need to consider the following elliptic curve:

$$tu^2 = -3\left(1 + 4k^3\right).$$

After calculation, we find that the elliptic curves $tu^2 = 1 + 4k^3$ and $tu^2 = -3(1 + 4k^3)$ have the same j-invariant, so they are isomorphic and have the same rank. If their ranks are greater than 1, both elliptic curves have infinitely many rational points. If their rank is equal to 0, we cannot distinguish their deflection points, and they may or may not have rational points. For $-50 \leq t \leq 50$ squarefree there are no other t matching the above solution except $t = -3$. Therefore, we propose a question.

Problem.2.1 Is there any other $t \neq -3$ such that (2.4) has non-zero solutions in $Q(\sqrt{t})$ for $n = 3$?

2.4   A New Attempt

After the famous Mordell conjecture was proved by the German mathematician Gerd Faltings (1954–) in 1984 (so that now it is known as Faltings's theorem) we already knew that for every given set of p, q, r, there are at most finitely many solutions of the (F–C) equation with genus greater than 1 (elliptic curves are algebraic curves with genus 1).

In the year following this proof, the French mathematician Joseph Oesterlé and the English mathematician David Masser each independently formulated the so-called *abc* conjecture. For any positive integer n, define its radical rad(n) to be the product of different prime factors of n. Then one of the three forms of the *abc* conjecture is

abc **Conjecture.** For any positive number $\varepsilon > 0$, there exist at most finitely many sets of positive integers (a, b, c) such that when $a + b = c, (a, b) = 1$ is satisfied and

$$q(a, b, c) = \frac{\log c}{\log(\mathrm{rad}(abc))} > 1 + \varepsilon.$$

If the *abc* conjecture holds, a number of famous theorems and conjectures can be easily derived, including Roth's (1958 Fields Prize), Baker's (1970 Fields Prize), Mordell's conjecture (1974 Fields Prize), and Fermat's Last Theorem (1998 IMU Silver Plaque).

If the *abc* conjecture holds, we can also prove directly that there are only finitely many solutions to the (F-C) equation, in other words that the Fermat–Catalan conjecture holds. In fact, let $p \leq q \leq r$, if $r \geq 8$, then $\frac{1}{p} + \frac{1}{q} + \frac{1}{r} \leq \frac{1}{2} + \frac{1}{3} + \frac{1}{8} = \frac{23}{24}$; if $r \leq 7$, then $\frac{1}{p} + \frac{1}{q} + \frac{1}{r}$ takes only finitely many values, the largest of which is $\frac{1}{2} + \frac{1}{3} + \frac{1}{7} = \frac{41}{42} > \frac{23}{24}$. Thus, it always holds that, $\frac{1}{p} + \frac{1}{q} + \frac{1}{r} \leq \frac{41}{42}$. If there are relatively prime $\{a, b, c\}$ satisfying the (F – C) equation, then from $x < z^{\frac{r}{p}}, y < z^{\frac{r}{q}}$, we have

$$q\left(x^p, y^q, z^r\right) = \frac{\log z^r}{\log(\mathrm{rad}(x^p y^q z^r))} = \frac{r \log z}{\log(\mathrm{rad}(xyz))} > \frac{r \log z}{\log(z^{r/p+q/p+1})}$$

$$= \frac{r}{r/p + r/q + 1} = \frac{1}{1/p + 1/q + 1/r} \geq \frac{42}{41}.$$

Therefore, it follows from the abc conjecture, the (F-C) equation has at most finitely many sets of solutions.

It is obvious that the Beal equation is a special case of the Fermat–Catalan equation. Therefore, if the abc conjecture holds, it can be introduced that there are at most finitely many counterexamples of the Beal conjecture, but it does not follow from the Fermat–Catalan conjecture that the Beal conjecture can settled completely.

> "In 2014, Preda Mihailescu (1955–), a Romanian-Swiss mathematician and professor at the University of Göttingen, Germany, who proved the Catalan conjecture, published a review of the abc conjecture in the Newsletter of the European Mathematical Society (see Mihailescu 1). In the last section of this article, he devoted the title "Varieties" to the additive and multiplicative equations proposed by the author and gave examples of the new Waring problem and the new Fermat equation."

The author having dinner with Mihailescu in Bucharest in 2018.

Figure 2.4. Institute for Advanced Study, photograph by the author in Princeton.

During the author's visit to the United States in the fall of 2018, he was invited to report at several universities on work related to additive equations, including the new Waring Problem and the new Fermat Problem. In particular, while reporting at the Department of Mathematics at Princeton University, Shai Evra, a member of (Institute for Advanced Study IAS, Princeton) (Fig. 2.4), showed great interest in this problem, and in the following year we exchanged a number of letters in which Evra suggested and proposed specific methods to study the new Fermat equation, mainly along the lines of Wiles' proof of Fermat's grand theorem.

Without loss of generality, one can assume that exactly one of A and B is even and the other odd with $4|(A+1), 32|B$.

Step 1. Consider the case where n is a prime $p \geq 5$ in the exponential (NF) equation The case of a semi-stable elliptic curve E defines a representation of the following Galois group over the field of rational numbers

$$\rho(E, p) : \mathrm{Gal}(\overline{\mathbb{Q}}/\mathbb{Q}) \to GL_2(F_p).$$

By Ribet's theorem (which had played a key role in Wiles' proof), this is integrable and divergent on 2 and p ramifies.

Step 2. Use the Taniyama–Shimura conjecture on semi-stable elliptic curves proved by Wiles, where E is a modular curve over the field of rational numbers. Further, let f be the right second order N_E of the Hecke cusp point modular form, where $N_E = \operatorname{rad}(ABC)$.

Then there exists a Galois representation

$$\rho(\mathrm{f}, \mathrm{p}) : \operatorname{Gal}(\overline{\mathbb{Q}}/\mathbb{Q}) \to GL_2(F_p).$$

It is the same as $\rho(E, p)$ up to isomorphism, i.e.,

$$\rho(E, p) \cong \rho(f, p),$$

Step 3. From Step 2, $\rho(E, p) \cong \rho(f, p)$, and its order is N_E. Then, by Step 1 and Ribet's theorem, it follows that for any odd prime factor q of N_E, if $\rho(E, p)$ is the cusp form of order N_E, then $\rho(E, p)$ must be the cusp form of order N_E/q.

Repeated use of this gives $\rho(E, p)$ is a cusp form of order 2.

However, a cusp form of order 2 does not exist, contradiction. Therefore, (2.4) has no positive integer solution.

We expect Evra's method to succeed, but just as Wiles had many difficulties in proving Fermat's Last Theorem, the final solution of Conjecture 1 will require the favor of the Goddess of Fate. In 2020, Evra was awarded the Ramanujan Prize, frequently regarded as a weather vane for the Fields Medal. It is worth noting that the above results for the prime exponent $p \geq 5$ can be extended to general integers n, but because of the requirements of Ribet's theorem, it is necessary to exclude $n = 2^a 3^b$. As mentioned before, there are already some examples of solutions for $n = 4$, whereas solutions for $n = 9$ have not yet been found. Therefore, we have the following

Problem 2.2. Does equation (2.4) have a solution when $n = 9$?

Problem 2.3. When $n = 4$, $gcd(A, B, C) \not\equiv 1, 5, 7 \pmod 8$, what is the general solution of equation (2.4)?

Finally, we would like to say that both the Beal conjecture and the Fermat–Catalan conjecture are a more natural idea and extension of the original Fermat's Last Theorem, whereas the formulation of the new Fermat Problem requires a bit of an imaginative leap. On the other hand, as pointed out by Mihailescu, our new Fermat's Last Theorem (Conjecture 2.4) remains firm even under the assumption of the strong abc conjecture. He also pointed out that perhaps the additive equation is "yin" and the multiplicative equation is "yang" in the eyes of Cai, in which case the additive and multiplicative equation is the yin-yang equation.

Chapter 3

Euler's Conjecture

Read Euler, read Euler, he is the master of us all

— *Pierre Simon Laplace*

3.1 A Disproven Conjecture

Throughout the history of mathematics, certain conjectures that have been disproven or rejected nevertheless continue to play a role and promote the development of the subject, a phenomenon that exudes an endless charm. For example, in 1640, Fermat considered the numbers

$$F_n = 2^{2^n} + 1$$

which are called Fermat numbers, and in particular Fermat primes when such a number is prime. Fermat verified himself that F_n is prime for $n = 0, 1, 2, 3$, or 4 (these numbers respectively are $3, 5, 17, 257, 65537$). He confidently proposed that F_n is prime for every nonnegative integer n. It was worth mentioning that any two Fermat numbers are relatively prime, which property was discovered by Christian Goldbach, from which we can obtain a new proof that there exist infinitely many prime numbers.

This conjecture survived for nearly a century until 1732, when the Swiss mathematician Leonhard Euler (Fig. 3.1), who was living in St. Petersburg (Figs. 3.2, 3.3a and 3.3b), proved that F_5 is not a

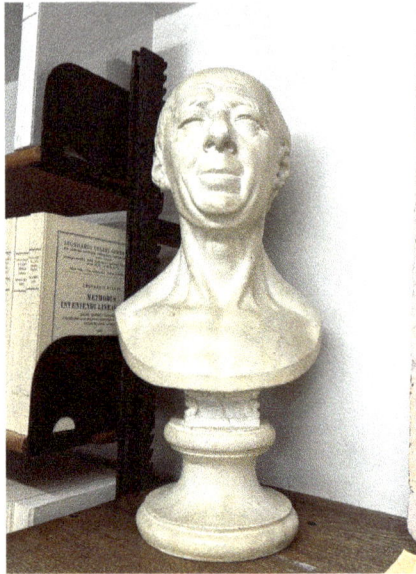

Figure 3.1. Portrait of Euler *changed to* Statue of Euler, photograph by the author in the University of Basel.

Figure 3.2. St. Petersburg Academy of Sciences, where Euler worked.

(a)

(b)

Figure 3.3. (a) Martin Church, where Euler was baptized, photograph by the author in Basel. (b) The Tomb of Euler, photograph by the author in St. Petersburg.

prime number, when he was younger than 25 years old; in fact, Euler proved that $641 \mid F_5$, using the following method:

Let $a = 2^7$, $b = 5$. Then $a - b^3 = 3$, $1 + ab - b^4 = 1 + 3b = 2^4$, so

$$F_5 = (2a)^4 + 1 = \left(1 + ab - b^4\right) a^4 + 1 = (1 + ab)a^4 + 1 - a^4 b^4$$

$$= (1 + ab)\{a^4 + (1 - ab)(1 + a^2 b^2)\}$$

where $1 + ab = 641$, $\quad F_5 = 4294967297 = 641 \times 6700417$.

Since then, mathematicians and enthusiasts have checked more than 40 numbers n, with the result that none of them produced prime numbers, including the integers $5 \leq n \leq 32$. So far, only the composite Fermat numbers corresponding to $n \leq 11$ have been fully factored; the smallest Fermat numbers for which we do not know any of the prime factors are F_{20} and F_{24}.

The factorizations of F_6, F_7, and F_8 (each of which has only two prime factors) were discovered by unknown mathematicians and published in 1880, 1971, and 1980 respectively. In 1877 and 1888, respectively, the Russian mathematician Ivan Pervushin (1827–1900), who also found the ninth Mersenne prime number, and lived to the east of the Ural Mountains, found a prime factor for each of F_{12} and F_{23}; these are

$$7 \times 2^{14} + 1 = 114689,$$

$$5 \times 2^{25} + 1 = 167772161.$$

The prime factor for F_{12} was also found by the French mathematician Edouard Lucas (1842–1891) in the same year.

The largest known composite Fermat number is $F_{3329780}$, which has a prime factor

$$193 \times 2^{3329782} + 1$$

which was discovered in July 2014. On the other hand, no new Fermat prime has been found in nearly three centuries.

There are many mysteries surrounding the Fermat numbers; for example, when $n \geq 5$, is F_n always composite? Or are there infinitely many Fermat primes? The latter problem was posed by the German mathematician Ferdinand Eisenstein (1823–1852) in 1844. Now that Fermat's Last Theorem has successfully been resolved, the problem of Fermat primes can be said to be the new last theorem of Fermat.

Since there are Fermat numbers, we must also have generalized Fermat numbers, which are defined to be the numbers $a^{2^n}+1$, with a any even number. Similarly, nobody knows whether or not infinitely many of the generalized Fermat numbers are prime or composite. In another direction, it has been found that the first six numbers of the form $2^{2^n}+15$ (that is, for $0 \leq n \leq 5$) are all prime These are respectively $17, 19, 31, 271, 65551, 4294967391$.

In 2014, the author introduced the GM-numbers, which are positive integers s satisfying

$$s = 2^\alpha + t$$

where t is the sum of the proper factors of s. Their name is derived from the Colombian author Gabriel García Márquez (1927–2014), who passed away just as the author was considering these numbers. His masterpiece, *One Hundred Years of Solitude*, reminds us of those mathematical problems that have remained unaddressed for centuries. It is obvious that when s is an odd prime, $t = 1$ and a must be a power of 2. In other words, odd prime GM-numbers are the same thing as Fermat primes. Moreover, the netizen Alpha helped to find two composite GM-numbers; these are

$$19649 = 7^2 \times 401 = 2^{14} + 3265,$$

$$22075325 = 5^2 \times 883013 = 2^{24} + 5298109.$$

Subsequently a search was carried out on integers not exceeding 2×10^{10}, finding only these two odd composite GM-numbers and the Fermat primes, for a total of 7 odd GM-numbers. As for even GM-numbers, they are much more common; there already six smaller than 100. These are $10, 14, 22, 38, 44, 92$, which can be seen from the presentations

$$10 = 2 + 8, \quad 14 = 2^2 + 10, \quad 22 = 2^3 + 14,$$

$$38 = 2^4 + 22, \quad 44 = 2^2 + 40, \quad 92 = 2^4 + 76.$$

Searching among integers not exceeding 10^6 and 10^8, we found respectively 146 and 350 even GM-numbers. We pose the following:

Question 3.1. Does there exist an eighth odd GM-number?

Question 3.2. Are there infinitely many even GM-numbers?

Supposing next that p is a prime number of the form $2^\alpha + 3$, $\alpha \geq 3$, then $2p$ must be a GM-number. This is because the sum of the proper factors of $2p$ is $p+3$, so $2p = 2^\alpha + p + 3$ has the required form. More generally, if a prime number p has the form $2^\alpha + 2^\beta - 1, \alpha \geq 1, \beta \geq 1$, then $2^{\beta-1}p$ is a GM-number. We make the following conjecture.

Conjecture 3.1. *There exist infinitely many positive integers α, β such that $2^\alpha + 2^\beta - 1$ is a prime number.*

It is easy to see that Conjecture 3.1 implies a positive answer to Question 3.2, that is, that there exist infinitely many even GM-numbers,

On the other hand, considering binary representations, every odd prime can be uniquely written as

$$1 + 2^{n_1} + \cdots + 2^{n_k} (1 \leq n_1 < \cdots < n_k).$$

We inquire whether, for a fixed given positive integer k, there are infinitely many prime numbers that can be expressed in this way. This is a generalization of the Eisenstein problem ($k = 1$). For any given prime number, we can assign it an order according to the value of k in the above representation. The order 1 prime numbers are 2 and the Fermat primes; the order 2 prime numbers include $7, 11, 13, 19, 41, \ldots$, the order 3 prime numbers include $23, 29, 43, 53, \ldots$, and so on.

Instead, if we set

$$t = \sum_{i=1}^{k} n_i$$

then we can classify the odd prime numbers according to the value of t; in this case, the number of elements in each class is finite. For example, the classes for $t = 1, 2, 3$ each have a single elements, respectively 3,5,7; the classes for $t = 4, 5$ both have two elements, respectively 11, 17, and 13, 19; the class for $t = 6$ is empty; the classes for $t = 7, 8$ both have three elements, respectively $23, 37, 67$ and $41, 131, 257$. We ask:

Question 3.3. With the exception of $t = 6$, is every such class of prime numbers non-empty?

Four years after Fermat introduced the Fermat numbers, his fellow mathematician Marin Mersenne (1588–1648), a Catholic priest in

Paris, proposed another problem, involving numbers which later generations have come to refer to as Mersenne numbers. These are numbers of the form

$$M_p = 2^p - 1$$

where p is a prime number. If M_p is also a prime number, then it is called a Mersenne prime; if instead M_p is composite, then it is called a composite Mersenne number.

In comparison with the Fermat primes, there are more Mersenne primes: so far, 51 of them have been found, and, following advances in computer technology, new ones are discovered every few years. Even more composite Mersenne numbers are known, but as with Fermat primes and composite Fermat numbers, we do not know whether or not there exist infinitely many Mersenne primes or composite Mersenne numbers. Intriguingly, the Mersenne primes are closely related to the ancient problem of perfect numbers. In addition to Mersenne, Leibniz and Goldbach also mistakenly believed that every Mersenne number is prime.

In the previous chapter, we discussed Fermat's Last Theorem and some variations. We turn now to Euler's conjecture, which is related to it. In 1769, when Euler had returned from St. Petersburg to Berlin, he sought to generalize Fermat's last theorem to multiple variables. His conjecture was as follows.

Euler's Conjecture. For every integer $s \geq 3$, the equation

$$a_1^s + a_2^s + \cdots + a_{s-1}^s = a_s^s \tag{3.1}$$

has no solutions in positive integers.

When $s = 3$, this is a special case of Fermat's Last Theorem, so naturally it is valid. For $s \geq 4$, the situation is different.

In 1911, R. Norrie discovered (see Norrie, 1911) that the equation

$$a^4 + b^4 + c^4 + d^4 = e^4 \tag{3.2}$$

has a solution given by $(a, b, c, d, e) = (30, 120, 272, 315, 353)$. This is not a counterexample to Euler's conjecture, in fact it is rather in line with his predictions.

But in 1966, almost three centuries after Euler first proposed his conjecture, L.J. Lander and T.R. Parkin carried out a computer search to obtain the first counterexample, for the case $s = 5$ (see Lander-Parkin, 1966):

$$27^5 + 84^5 + 110^5 + 133^5 = 144^5.$$

Unfortunately, similar such searches for the case $s = 4$ have proved fruitless.

In 1988, Noam Elkies, at that time a student at Harvard University (Fig. 3.4) used (see Elkies, 1988) the theory of elliptic curves to obtain infinitely many counterexamples to Euler's conjecture in the case $s = 4$; one of these, for example, is

$$2682440^4 + 15365639^4 + 18796760^4 = 20615673^4.$$

In the same year, Roger Frye, working for Thinking Machines Corporation, provided a smaller counterexample, in fact the smallest counterexample to Euler's conjecture for $s = 4$, which is

$$95800^4 + 217519^4 + 414560^4 = 422481^4.$$

Figure 3.4. The Mathematics Building of Harvard University.

In 2004, Jim Frye found a counterexample for the case $s = 5$:

$$55^5 + 3183^5 + 28969^5 + 85282^5 = 85359^5.$$

For $s \geq 6$, there is as of yet neither any counterexample nor any proof for Euler's conjecture.

In 2008, the physicist Lee W. Jacobi and mathematician Daniel J. Madden, both American, studied the nonzero integer solutions of the $s = 4$ Euler equation with linear term restrictions, namely:

$$a^4 + b^4 + c^4 + d^4 = (a + b + c + d)^4 \tag{3.3}$$

by transforming it into the equivalent Pythagorean triple

$$(a^2 + ab + b^2)^2 + (c^2 + cd + d^2)^2 = ((a+b)^2 + (a+b)(c+d) + (c+d)^2)^2.$$

Using two solutions

$$(a, b, c, d) = (955, -2634, 1770, 5400), (7590, -31764, 27385, 48150)$$

discovered around 1964 by S. Brudno and J. Wróblewski and the theory of elliptic curves, they proved that (3.3) has infinitely many solutions in nonzero integers, and therefore so does (3.2).

It is worth mentioning that counterexamples to Euler's conjecture have also aroused the interest of geometers; for example, in the discipline of Diophantine geometry on number fields, there is the so-called Skorobogotov conjecture, which is related to it.

Continuing on, we first describe the proof due to Elkies, and then the work of Jacobi and Madden, before finally introducing a generalization of Euler's conjecture via the additive and multiplication method.

3.2 The Infinitely Many Counterexamples of Elkies

Considering equation (3.1), when $s = 3$, it is equivalent to the following equation:

$$r^4 + s^4 + t^4 = 1 \tag{3.4}$$

where r, s, t are rational numbers. We first look at the equation

$$r^4 + s^4 + t^2 = 1 \tag{3.5}$$

In 1973, V. A. Dem'janenko transformed (see Dem'janenko, 1973/74) the above equation into the following family of conic sections in parameter u:

$$r = x + y, \quad s = x - y \tag{3.6a}$$

$$\left(u^2 + 2\right) y^2 = -(3u^2 - 8u + 6)x^2 - 2(u^2 - 2)x - 2u \tag{3.6b}$$

$$\left(u^2 + 2\right) t = 4 \left(u^2 - 2\right) x^2 + 8ux + \left(2 - u^2\right). \tag{3.6c}$$

Subsequently, A. Bremner, Don Zagier, and Elkies each used different methods to rederive (3.6). Bremner noticed that (3.4) is equivalent to the following identity, and then factored both sides of the equation in $Q(\sqrt{-1})$:

$$2(1 + r^2)(1 + s^2) = (1 + r^2 + s^2)^2 + t^2.$$

Using the relationship between roots and coefficients, from (3.6b) we can obtain,

$$\begin{aligned}
u &= \frac{-1 + 4x^2 \pm \sqrt{1 - (2x^4 + 12x^2y^2 + 2y^4)}}{3x^2 + y^2 + 2x} \\
&= \frac{-1 + (r + s)^2 \pm \sqrt{1 - r^4 - s^4}}{r^2 + rs + s^2 + r + s} \\
&= \frac{-1 + (r + s)^2 \pm t}{r^2 + rs + s^2 + r + s};
\end{aligned} \tag{3.7a}$$

Therefore, u is a rational number, say $u = 2m/n$, with $m \geq 0, n$ odd, and $(m, n) = 1$; we can replace u with $2/u$, so that (3.6b) and (3.6c) can be written as

$$(2m^2 + n^2)y^2 = -(6m^2 - 8mn + 3n^2)x^2 - 2\left(2m^2 - n^2\right)x - 2mn, \tag{3.7b}$$

$$(2m^2 + n^2)t = 4(2m^2 - n^2)x^2 + 8mnx + (n^2 - 2m^2). \tag{3.7c}$$

If (3.7b) has a rational point (x, y), we can find t from (3.7c), and r and s from (3.6a), giving a solution to (3.5). So we need to solve (3.7b), and for this purpose we introduce the following lemma.

Lemma 3.1. *The conic section* (3.7b) *has infinitely many rational points if and only if* $R(2m^2 + n^2)$ *and* $R(2m^2 - mn + n^2)$ *only have* *prime factors congruent to 1 modulo 8, where* $R(\mathrm{k})$ *is defined as follows: for any nonnegative integer* k, *let* $S(k)$ *be the largest positive integer with square dividing* k, *and let*

$$R(k) = k/S^2(k).$$

For example, for $k = \pm 23, \pm 24, \pm 25$, we have $S(k) = 1, 2, 5$ and $R(k) = \pm 23, \pm 6, \pm 1$ respectively.

First, we work out an example. When $u = 4, (m, n) = (2, 1)$, satisfies the conditions of Lemma 3.1. Then (3.7b) becomes

$$9y^2 = -11x^2 - 14x - 4. \tag{3.8}$$

We observe that $(x, y) = \left(-\frac{1}{2}\frac{1}{6}\right)$ is a solution to (3.8), from which we can derive the solution $(r, s, t) = \left(\frac{1}{3}, \frac{2}{3}, \frac{8}{9}\right)$ to (3.6) or (3.5). Then starting from the solution $(x, y) = \left(-\frac{1}{2}, \frac{1}{6}\right)$, we can obtain by projection a family of solutions to (3.8) in parameter k:

$$(x, y) = \left(-\frac{k^2 + 2k + 17}{2k^2 + 22}, -\frac{k^2 + 6k - 11}{6k^2 + 66}\right).$$

Working backwards, we get an infinitely family of solutions to (3.5):

(r, s, t)

$$= \left(\frac{2k^2 + 6k + 20}{3k^2 + 33}, \frac{k^2 + 31}{3k^2 + 33}, \frac{4\left(2x^4 - 3k^3 + 28k^2 - 75k + 80\right)}{(3k^2 + 33)^2}\right).$$

In general, for any u satisfying the conditions of Lemma 3.1, we have an infinite family of solutions to (3.5).

Now, we can find solutions to (3.4) from solutions to (3.5); clearly, we need only that $\pm t$ is a square. As before, the solutions must satisfy

$$r = x + y, \quad s = x - y \tag{3.9a}$$

$$(u^2 + 2)y^2 = -(3u^2 - 8u + 6)x^2 - 2(u^2 - 2)x - 2u \tag{3.9b}$$

$$\pm(2m^2 + n^2)t^2 = 4(2m^2 - n^2)x^2 + 8mnx + (n^2 - 2m^2). \tag{3.9c}$$

Note that the left-hand side of (3.9c) here is slightly different. To solve (3.9c) we need another lemma, which is similar to Lemma 3.1.

Lemma 3.2. *The conic section* (3.9c) *has infinitely many solutions if and only if all the prime factors of* $R(2m^2 - 2mn + n^2)$, $R(2m^2 + n^2)$ $(2m^2 + n^2)$ *and* $R(2m^2 + 2mn + n^2)$ *are congruent to 1 modulo 8.*

The proofs of Lemmas 3.1 and 3.2 given by Elkies are elementary, but not short, and we omit them here. From the proof of Lemma 3.2, it turns out that m must be a multiple of 4. It is not hard to find the smallest solutions $(m, n) = (0, 1), (4, -7), (8, -5), (8, -15), (12, 5), (20, -1)$ and $(20, -9)$ satisfying these two lemmas.

When $(m, n) = (1, 0)$, the corresponding solution to (3.4) is obviously $(\pm 1, 0, 0)$, while $(m, n) = (4, -7)$ does not produce any solution to (3.4). Considering $(m, n) = (8, -5)$, and substituting in (3.7b) and (3.9c), we get

$$153y^2 = -779x^2 - 206x + 80$$

$$\pm 153y^2 = 412x^2 - 320x - 103.$$

For the first of the conic sections, after multiple trials, we find a smaller solution $(x, y) = (3/14, 1/42)$. This gives the parameterized family

$$(x, y) = \left(\frac{51k^2 - 34k - 5221}{14(17k^2 + 779)}, \frac{17k^2 + 7558k - 779}{42(17k^2 + 779)} \right).$$

Substituting this value of x into the second of the conic sections and simplifying, we get

$$\pm 21^2 \left(17k^2 + 779 \right)^2 t^2$$

$$= -4 \left(31790k^4 - 4267k^3 + 1963180k^2 - 974003k - 63237532 \right). \tag{3.10}$$

Taking the positive value on the left hand side of the above equation, we have the transformation

$$X = (k + 2)/7, \quad Y = 3(17k^2 + 779t)/14$$

and (3.10) becomes

$$Y^2 = -31790X^4 + 36941X^3 - 56158X^2 + 28849X + 22030. \tag{3.11}$$

For (3.11), Elkies found by computer search the solution

$$(X, Y) = \left(-\frac{31}{467}, \frac{30731278}{467^2} \right).$$

Working backwards, we have as a solution for (3.4)

$$(r, s, t) = \left(-\frac{18796760}{20615673}, \frac{2682440}{20615673}, \frac{15365639}{20615673} \right).$$

Clearing the denominator provides a counterexample to the $s = 4$ case of Euler's conjecture:

$$2682440^4 + 15365639^4 + 18796760^4 = 20615673^4.$$

In fact, any number of solutions can be derived from (3.11) according to the following.

Proposition 3.1. *There are infinitely many X such that the right-hand side*

$$Y^2 = -31790X^4 + 36941X^3 - 56158X^2 + 28849X + 22030$$

of (3.11) is a square; in particular, (3.4) has infinitely many solutions.

The proofs of Proposition 3.1 and, in the next section, Theorem 3.1 both rely on the famous Mazur's theorem on torsion of elliptic curves (see Silverman-Tate, 2004, p. 58).

Mazur's Theorem. *Elliptic curves have at most 16 rational torsion points (points of finite order).*

According to Mazur's theorem, an elliptic curve with at least 17 distinct rational points must have infinitely many rational points.

Proof of Proposition 3.1. Given the two rational points

$$P_\pm \colon (X, Y) = \left(-\frac{31}{467}, \pm\frac{30731278}{467^2} \right)$$

on the elliptic curve (3.10), we need to show that their difference $Q = P_+ - P_-$ is of infinite order in its Jacobian group; that is to say, it is not a torsion point. From Mazur's theorem, there are at most finitely many rational torsion subgroups on the elliptic curve, and in particular no torsion point has index exceeding 12, so the proof of the proposition reduces to a finite computation; specifically, we verify that $n \cdot Q \neq 0$ for each of $n = 2, 3, \ldots, 12$.

The number of calculations can be reduced even further by the observation that the Jacobi group of the elliptic curve (3.11) has an order 2 rational point, corresponding to (3.9) via $(x, y, t) \leftrightarrow (x, -y, -t)$. So we need only to check that $n \cdot Q$ is neither zero nor an order 2 point for $n = 2, 3, \ldots, 6$, which can be verified by calculation, completing the proof of Proposition 3.1.

Elkies proved moreover that the rational solutions of (3.4) are dense in the space of all real solutions to (3.4).

When R. Frye learned of this $n = 4$ counterexamples, he asked Elkies if this solution included the smallest possible counterexample; Elkies recommended to consider the following conditions: take D an odd integer, not divisible by $5, C < D$ such that 625 divides $(D^4 - C^4)$, along with some other specified congruences. In that year, Frye spent a hundred hours using a computer network to obtain the smallest counterexample.

Subsequently, Frye also checked by computer that his own counterexample was the only one in the range $D < 10^6$; however it remains unknown whether the second smallest counterexample is the one due to Elkies, corresponding to $(m, n) = (20, -9)$ in (3.9), discussed above, and satisfying the conditions of Lemmas 3.1 and 3.2

3.3 The Euler Equation with Linear Constraints

In 1964 S. Brudno found another solution to (3.2):

$$5400^4 + 1770^4 + 2634^4 + 955^4 = 5491^4.$$

He noticed that among these numbers

$$5400 + 1770 + 955 = 2634 + 5491$$

or in other words $(5400, 1770, -2634, 955)$ is a solution to the following Euler equation with a linear constraint

$$a^4 + b^4 + c^4 + d^4 = (a + b + c + d)^4. \tag{3.12}$$

In 2008, Lee W. Jacobi and Daniel J. Madden proved (see Jacobi-Madden, 2008):

Theorem 3.1. *Equation* (3.12) *has infinitely many solutions in rational numbers.*

In order to prove Theorem 3.1, we invoke the following lemma.

Lemma 3.3. *Suppose* (x_1, y_1, z_1) *is a solution to the following equation:*

$$z^2 y^2 = \alpha_4 x^4 + \alpha_3 x^3 z + \alpha_2 x^2 z^2 + \alpha_1 x z^3 \alpha_0 z^4.$$

Then the above equation also has a solution given by

$$x_2 = \left(64 x_1 y_1^6 z_1^6 y_4 - q_0^2 x_1 - 64 y_1^6 z_1^6 y_3 + 8 q_0 y_1^2 z_1^2 y_1\right)\left(64 y_1^6 z_1^6 y_4 - q_0^2\right)$$

$$y_2 = 8 q_0 y_1 \left(q_0 y_1 - 8 y_1^4 z_1^4 y_3\right)^2 + 4 y_1 y_1 \left(q_0 y_1 - 8 y_1^4 z_1^4 \gamma_3\right)$$

$$\times \left(64 y_1^6 z_1^6 y_4 - q_0^2\right) + y_1 \left(64 y_1^6 z_1^6 y_4 - q_0^2\right)^2$$

$$z_2 = z_1 \left(64 y_1^6 z_1^6 y_4 - q_0^2\right)^2$$

where

$$\gamma_1 = 4 \alpha_4 x_1^3 + 3 \alpha_3 x_1^2 z_1 + 2 \alpha_2 x_1 z_1^2 + \alpha_1 z_1^3$$

$$\gamma_2 = 6 \alpha_4 x_1^2 + 3 \alpha_3 x_1 z_1 + \alpha_2 z_1^2$$

$$\gamma_3 = 4 \alpha_4 x_1 + \alpha_3 z_1$$

$$\gamma_4 = \alpha_4$$

$$q_0 = 4 y_1^2 z_1^2 \gamma_2 - \gamma_1^2.$$

Proof of Theorem 3.1. First, we rewrite (3.12) as

$$a^4 + b^4 + (a+b)^4 + c^4 + d^4 + (c+d)^4$$

$$= (a+b)^4 + (c+d)^4 + (a+b+c+d)^4.$$

Using the identity

$$\alpha^4 + \beta^4 + (\alpha + \beta)^4 = 2\left(\alpha^2 + \alpha\beta + \beta^2\right)^2$$

we can make the further transformation

$$\left(a^2 + ab + b^2\right)^2 + \left(c^2 + cd + d^2\right)^2$$

$$= \left((a+b)^2 + (a+b)(c+d) + (c+d)^2\right)^2$$

or after rearranging terms

$$\left(c^2 + cd + d^2\right)^2$$
$$= \left((a+b)^2 + (a+b)(c+d) + (c+d)^2\right)^2 - \left(a^2 + ab + b^2\right)^2$$
$$= \left((a+b)^2 + (a+b)(c+d) + (c+d)^2 + a^2 + ab + b^2\right)$$
$$\left((a+b)^2 + (a+b)(c+d) + (c+d)^2 - a^2 - ab - b^2\right).$$

Introducing the parameter μ, we have

$$c^2 + cd + d^2 = \mu((a+b)^2 + (a+b)(c+d) + (c+d)^2 - a^2 - ab - b^2)$$

$$c^2 + cd + d^2 = \frac{1}{\mu}((a+b)^2 + (a+b)(c+d) + (c+d)^2 + a^2 + ab + b^2).$$

Starting from the known solution $(a, b, c, d) = (5400, 1770, -2634,$ $955)$, we can determine the value of the parameter of the above quadratic surface as

$$\mu_0 = \frac{961}{61}.$$

At the same time, the equation of this quadratic surface can be written as

$$c^2 + cd + d^2 = \mu_0(ab + ac + bc + ad + bd + c^2 + 2cd + d^2)$$

$$\frac{1}{\mu_0}(c^2 + cd + d^2) = 2a^2 + 3ab + 2b^2$$

$$+ ac + bc + ad + bd + c^2 + 2cd + d^2.$$

We would like for the intersection of these two quadratic surfaces to be an elliptic curve; we have the transformation

$$\begin{pmatrix} a \\ b \\ c \\ d \end{pmatrix} = \begin{pmatrix} 0 & 0 & 2 & 2 \\ 0 & 0 & 2 & -2 \\ -1 & -1 & -1 & 0 \\ 1 & -1 & -1 & 0 \end{pmatrix} \begin{pmatrix} x \\ y \\ z \\ w \end{pmatrix}$$

under which the quadratic surfaces become

$$61(x^2 + 3y^2 + 6yz + 3z^2) = 961(4y^2 - 4w^2)$$
$$961(x^2 + 3y^2 + 6yz + 3z^2) = 61(4w^2 + 4y^2 + 24z^2).$$

We move the right-hand side of the first equation to the left and subtract the second from it to get

$$61x^2 - 3661y^2 + 366yz + 183z^2 + 3844w^2 = 0, \qquad (3.13a)$$

$$459900y^2 - 11163z^2 - 463621w^2 = 0. \qquad (3.13b)$$

This last equation (3.13b) is a conic section. According to these transformations, Brudno's solution is equivalent to

$$(x_0, y_0, z_0, w_0) = \left(\frac{3589}{2}, -953, \frac{3585}{2}, \frac{1815}{2} \right).$$

This solution gives us a parametric solution to (3.16b):

$$y_1 = -6(146094900t^2 + 13339785st + 3546113s^2)$$

$$z_1 = 45(36638700t^2 + 38958640st + 889319s^2)$$

$$w_1 = 5445(153300t^2 - 3721s^2)$$

with which we would like to determine the solutions of (3.13a), in other words to obtain

$$(3661y^2 - 366yz - 183z^2 - 3844w^2)/61$$

as a square number. That is, we need to find s, t such that

$$1605124656896049s^4 + 26478277616573460s^3t$$

$$- 3598879905807952500^2t^2 + 1090868035103658000st^3$$

$$+ 2724417967677210000^4$$

is square.

The situation is trivial when either $s = 0$ or $t = 0$; in order to avoid these trivial cases, we consider

$$x^2t^2 = 1605124656896049s^4 + 26478277616573460s^3t$$

$$- 3598879905807952500s^2t^2 + 1090868035103658000st^3$$

$$+ 272441796767721000t^4. \qquad (3.14)$$

This is a homogeneous equation of an elliptic curve, and every rational point on it produces a solution to (3.12), as expressed by the following formulas:

$$a = 39517020s^2 + 3506277600st + 49669200000t^2$$

$$b = 120560400s^2 + 3506277600st + 1628046000t^2$$

$$c = -18742677s^2 - 1637100090st - 772172100t^2 - tx$$

$$d = -18742677s^2 - 1637100090st - 772172100t^2 + tx.$$

In particular, two rational points on the curve (3.14) are given by $(s, t, x) = (0, 1, \pm1650581100)$, containing the solution due to Brudno. From these rational points, it is possible to determine others by arithmetic operations, but their values grow very fast, prompting Jacobi and Madden to prove directly that there exist infinitely many of them. To this end, they worked modulo the carefully selected prime number 71 and identified 18 distinct rational points.

Starting from the rational point $(s : x : t) = (0, 1650581100, 1)$ of (3.14) and repeatedly invoking Lemma 3.3, we obtain a sequence of rational points along the curve, all of them with integer coordinates; working modulo 71, we have the following 18 distinct points.

s	0	53	44	41	47	60	15	39	2
x	±9	±64	±31	±1	±8	±48	±41	±56	±10
t	1	50	24	1	54	8	10	6	15

We divide through by t modulo 71 by defining

$$\frac{s}{t} \equiv s'(71), \quad \frac{x}{t} \equiv x'(71)$$

and obtain in this way the following 18 points.

s	0	11	61	41	39	43	37	42	38
x	±9	±24	±22	±1	±58	±6	±68	±33	±48
t	1	1	1	1	1	1	1	1	1

These 18 points $(s: x : 1)$ are all different modulo 71; invoking Mazur's theorem, if the elliptic curve contains more than 16 rational points, then it has infinitely many. In other words, there are infinitely many primitive solutions for equation (3.12), proving Theorem 3.1.

3.4   Variants of Euler's Conjecture

In April of 2012, inspired by the new Waring problem, the author proposed to solve the following system of Diophantine equations

$$\begin{cases} n = a_1 + a_2 + \cdots + a_{s-1} \\ a_1 a_2 \cdots a_{s-1} (a_1 + a_2 + \cdots + a_{s-1}) = b^s \end{cases} \tag{3.15}$$

where $s \geq 3, n, a_i, b$ are all positive integers.

This is another type of additive and multiplicative equation. It is clear that a solution to (3.15) can necessarily be obtained from a solution to (3.1), by taking

$$n = a_s^s, a_1 = a_1^s, \ldots, a_{s-1} = a_{s-1}^s.$$

We (Cai Tianxin, Zhang Yong) tried to find a counterexample to Euler's conjecture ($n = 6$) via (3.15). Although we did not succeed, we nevertheless obtained several beautiful results using the theory of elliptic curves (see Cai and Zhang, 2021a or Cai and Zhang, 2021b).

Theorem 3.2. *When $s = 3$ and n is any positive integer, there are no solutions to (3.15).*

Proof.   When $s = 3$, (3.15) is

$$\begin{cases} n = a_1 + a_2 \\ a_1 a_2 (a_1 + a_2) = b^3. \end{cases} \tag{3.16}$$

From the second formula, we find

$$\frac{a_1}{b} \frac{a_2}{b} \left(\frac{a_1}{b} + \frac{a_2}{b} \right) = 1.$$

Set $b_i = \frac{a_i}{b} (1 \leq i \leq 2)$; then this is

$$b_1 b_2 (b_1 + b_2) = 1$$

so that

$$\left(\frac{b_1}{b_2} \right)^2 + \frac{b_1}{b_2} = \frac{1}{b_2^3}.$$

Substituting again $u = \frac{b_1}{b_2}, v = \frac{1}{b_2}$, we get

$$u^2 + u = v^3$$

Setting $y = 8u + 4, x = 4v$, we get

$$y^2 = x^3 + 16.$$

Then using the Magma computational package, we find that the only rational points are given by the trivial solutions

$$(x, y) = (0, \pm 4)$$

from which it follows that (3.16) has no integer solutions, proving Theorem 3.2. □

According to Theorem 3.2, Fermat's Last Theorem holds for $n = 3$, this is much simpler than Euler's original proof.

Theorem 3.3. *When $s = 4$, there are infinitely many positive integers n such that (3.15) has infinitely many solutions in positive integers.*

Proof. When $s = 4$, (3.15) is

$$\begin{cases} n = a_1 + a_2 + a_3 \\ a_1 a_2 a_3 (a_1 + a_2 + a_3) = b^4. \end{cases} \tag{3.17}$$
□

From the second equation, we get

$$\frac{a_1}{b} \frac{a_2}{b} \frac{a_3}{b} \left(\frac{a_1}{b} + \frac{a_2}{b} + \frac{a_3}{b} \right) = 1.$$

Putting $b_i = \frac{a_i}{b} (1 \le i \le 3)$, this is

$$b_1 b_2 b_3 (b_1 + b_2 + b_3) = 1.$$

By inspection, $(a_1, a_2, a_3) = (1, 2, 24), b = 6$ is a solution of (3.17), which gives

$$(b_1, b_2, b_3) = \left(\frac{1}{6}, \frac{1}{3}, 4 \right)$$

and the corresponding identities

$$\begin{cases} b_1 b_2 b_3 = \frac{2}{9} \\ b_1 + b_2 + b_3 = \frac{9}{2}. \end{cases} \tag{3.18}$$

Next we view b_i as an unknown. Eliminating b_3 from (3.18), we get

$$18 b_1^2 b_2 + 18 b_1 b_2^2 - 81 b_1 b_2 + 4 = 0.$$

Then with $u = \frac{b_1}{b_2}$, $v = \frac{1}{b_1}$, this is

$$18u^2 + 18u - 81u + 4v^3 = 0.$$

Finally, setting

$$y = 384u - 864v + 192, x = -32v + 243$$

we obtain the equation of an elliptic curve

$$E : y^2 = x^3 - 166779x + 26215254.$$

By the Nagell–Lutz theorem, it suffices to show that E contains infinitely many rational points to find a single rational point on E the x-coordinate of which is not an integer; using Magma, we find the rational point $\left(\frac{30507}{121}, -\frac{584592}{1331}\right)$ on E.

From the above transformation we have

$$\begin{cases} b_1 = \frac{32}{243-x} \\ b_2 = \frac{-y+27x-6369}{12(243-x)} \\ b_3 = \frac{y+27x-6369}{243-x}. \end{cases}$$

Therefore,

$$\begin{cases} a_1 = \frac{32}{243-x}b \\ a_2 = \frac{y-27x+6369}{12(243-x)}b \\ a_3 = \frac{-y-27x+6369}{243-x}b \end{cases}$$

is a solution to (3.17).

In order that every $b_i > 0$, it is necessary to have

$$x < 243, \quad |y| < 27x - 6369.$$

Now, by the Poincaré–Hurwitz theorem, there are infinitely many rational points of E in any neighborhood of any of its rational points, so we need only to find a single point satisfying the above constraints; clearly the point $P = (235, 8)$ works, so we conclude that there are infinitely many rational points (x, y) satisfying the above constraints.

It follows that we can find infinitely many rational solutions $b_i > 0 (1 \le i \le 3)$ to (3.18), and clearing the denominators from the b_i, we obtain infinitely many integer points $a_i > 0$. This completes the proof that for $s = 4$, equation (3.15) has infinitely many solutions in positive integers.

Example 3.1. With $s = 4$, the rational points

$$(x, y) = (235, 8), \left(\frac{60266587}{257049}, \frac{3852230624}{130323843} \right)$$

on the elliptic curve E give the solutions

$$(a_1, a_2, a_3) = (1, 2, 24), (781943058, 138991832, 18609625)$$

to (3.18), corresponding to

$$\begin{cases} 27 = 1 + 2 + 24 \\ 1 \cdot 2 \cdot 24 \cdot (1 + 2 + 24) = 6^4 \end{cases}$$

and

$$\begin{cases} 939544515 = 781943058 + 138991832 + 18609625 \\ 781943058 \cdot 1389918321 \cdot 18609625 \\ \quad \times (781943058 + 138991832 + 18609625) = 208787670^4. \end{cases}$$

Theorem 3.4. *For $s \ge 5$, there exist infinitely many positive integers n such that (3.15) has infinitely many solutions in positive integers; more precisely, there are infinitely many positive integers n such that (315) has infinitely many solutions in $s - 3$ parameters.*

Proof. For each $1 \le i \le s - 1$, put

$$b_i = \frac{a_i}{b} \in Q^+.$$

Then (3.15) can be simplified to

$$\begin{cases} \frac{n}{b} = b_1 + b_2 + \cdots + b_{s-1} \\ b_1 b_2 \cdots b_{s-1} (b_1 + b_2 + \cdots + b_{s-1}) = 1. \end{cases}$$

Making the substitutions

$$x = b_1, y = b_2, z = b_3, u = b_4 \cdots b_{s-1}, v = b_4 + \cdots + b_{s-1}$$

we have

$$\begin{cases} \frac{n}{b} = x + y + z + v \\ xyzu(x + y + z + v) = 1. \end{cases}$$

We investigate the rational points of the second of these equations; setting $z = ut^2y$, this becomes

$$t^2u^2y^2x^2 + t^2u^2y^2(t^2uy + v + y)x - 1 = 0.$$

Viewing this equation as a quadratic in x, the existence of a rational solution implies that the discriminant

$$\Delta(y) = u^2t^2y^2(u^2t^2(ut^2 + 1)^2y^4 + 2vu^2t^2(ut^2 + 1)y^3 + u^2v^2t^2y^2 + 4)$$

must be a square. We define the quartic curve C:

$$w^2 = u^2t^2\left(ut^2 + 1\right)^2 y^4 + 2vu^2t^2\left(ut^2 + 1\right)y^3 + u^2v^2t^2y^2 + 4.$$

It is easy to calculate that the discriminant of C is

$$\Delta(t) = 256u^6t^6(ut^2 + 1)^4(64u^2t^4 + u(uv^4 + 128)t^2 + 64).$$

Since u and v are positive rational numbers, the discriminant of C is smooth and non-zero. Transform C into the family of elliptic curves (see Cohen, 2007, Proposition 7.2.1)

$$E : Y^2 = X(X^2 + u^2v^2t^2X - 16u^2(ut^2 + 1)^2t^2).$$

Here, we make use of the birational map $\varphi : C \to E$ given by

$$y = \frac{Y - uvtX}{ut\left(ut^2 + 1\right)X}, \qquad w = \frac{Y^3 - u^2v^2t^2X^3 - 2X^3}{4ut\left(ut^2 + 1\right)X^3}$$

and

$$X = 2ut\left(ut^2ut^2 + 1\right)\left(t^3u^2y^2 + tuvy + tuy^2 - w\right)$$

$$Y = 2u^2t^2\left(ut^2 + 1\right)\left(2t^2uy + v + 2y\right)\left(t^3u^2y^2 + tuvy - tuy^2 - w\right).$$

Since u, v are positive and rational, we can write

$$u = \frac{p}{q}, \qquad v = \frac{c}{d}$$

where p, q, c, d are positive integers. Then let

$$U = q^4 d^2 X_v, \quad V = q^6 d^3 Y$$

and E can be converted to

$$E' : V^2 = U(U^2 + c^2 p^2 q^2 t^2 U - 16 d^4 p^2 q^4 t^2 (pt^2 + q)^2).$$

Noting that the point

$$P = (4pd^2 q^2 (pt^2 + q)t, 4cd^2 p^2 q^3 (pt^2 + q)t^2)$$

lies on E' and taking advantage of the group law on elliptic curves, we obtain the point

$$2P = \left(\frac{16 q^4 d^4 \left(pt^2 + q \right)^2}{c^3}, \frac{64 q^3 d^6 \left(pt^2 + q \right)^3}{c^3} \right).$$

We can now find the positive rational solutions of (3.15). Considering the reflection point $-2P$ of $2P$ under the bijection φ, we get

$$x = \frac{uv^3 t}{2 \left(4ut^2 - uv^2 t + 4 \right)},$$

$$y = \frac{4ut^2 - uv^2 t + 4}{2uvt \left(ut^2 + 1 \right)},$$

$$z = \frac{\left(4ut^2 - uv^2 t + 4 \right) t}{2v \left(ut^2 + 1 \right)}.$$

Since $u, v > 0$, in order to have $x, y, z > 0$, it is necessary that

$$4ut^2 - uv^2 t + 4 > 0 \tag{3.19}$$

a quadratic with discriminant $\delta = u \left(uv^4 - 64 \right)$. If $\delta < 0$, then (3.19) holds for any positive integer t. If $\delta > 0$, then when (3.19) holds when

$$t \in \left(0, \frac{uv^2 - \sqrt{u \left(uv^4 - 64 \right)}}{8u} \right) \cup \left(\frac{uv^2 + \sqrt{u \left(uv^4 - 64 \right)}}{8u}, \infty \right).$$

In either case when $u, v > 0$, there are infinitely many t such that $x, y, z > 0$. Therefore, for any given positive rational numbers

$b_4 + \cdots + b_{s-1}$, there exist infinitely many positive rational numbers t such that $b_i > 0, i = 1, 2, 3$. Clearing the denominators of $t, b_i (i = 1, \ldots, s-1)$ gives the solution $a_i (i = 1, \ldots, s-1)$ in positive integers.

Therefore, for $s \geq 5$, there are infinitely many n such that (3.15) has a solution $(a_1, a_2, \ldots, a_{s-1}, b)$ in positive integers and with $s - 3$ parameters $t, b_i (i = 4, \ldots, s - 1)$. This completes the proof of Theorem 3.4. $\qquad\square$

Corollary 3.1. *For $s \geq 4$ and any positive integer n, (3.15) admits a positive solution.*

This is because it follows from Theorems 3.3 and 3.4 that for $s \geq 4$ there exists a positive integer n such that (3.15) has positive rational solutions. Then for any positive integer N we need only to make the following transformations:

$$\begin{cases} N = \frac{a_1 N}{n} + \frac{a_2 N}{n} + \cdots + \frac{a_{s-1} N}{n} \\ \frac{a_1 N}{n} \cdot \frac{a_2 N}{n} \cdots \frac{a_{s-1} N}{n} \cdot \left(\frac{a_1 N}{n} + \frac{a_2 N}{n} + \cdots + \frac{a_{s-1} N}{n} \right) = \left(\frac{bN}{n} \right)^s. \end{cases}$$

Example 3.2. With $s = 5$, (3.15) has positive integer solutions $(2, 28, 49, 49)(b = 28), (5, 81, 90, 324)(b = 90)$.

In fact, we have

$$x = \frac{uv^3 t}{2 \left(4ut^2 - wv^2 t + 4 \right)}, \quad y = \frac{4ut^2 - uv^2 t + 4}{2uvt \left(ut^2 + 1 \right)},$$

$$z = \frac{\left(4ut^3 - uv^2 t + 4 \right) t}{2v \left(ut^2 + 1 \right)}, \quad u = v = b_4.$$

Set

$$b = 2uvt \left(ut^2 + 1 \right) \left(4ut^2 - uv^2 t + 44ut^2 - uv^2 t + 4 \right).$$

Then

$$a_1 = t^2 u^2 \left(ut^2 + 1 \right),$$
$$a_2 = \left(4ut^2 - tu^3 + 4 \right)^2,$$
$$a_3 = ut^2 \left(4ut^2 - tu^3 + 4 \right)^2,$$
$$a_4 = 2tu^3 \left(ut^2 + 1 \right) \left(4ut^2 - tu^3 + 4 \right)^2.$$

If $t = 1$, $u = 1$, we get

$$\begin{cases} 128 = 2 + 49 + 49 + 28 \\ 2 \cdot 49 \cdot 49 \cdot 28 \cdot (2 + 49 + 49 + 28) = 28^5. \end{cases}$$

If $t = 2, u = 1$, we get

$$\begin{cases} 2000 = 20 + 324 + 1296 + 360 \\ 20 \cdot 324 \cdot 1296 \cdot 360 \cdot (20 + 324 + 1296 + 360) = 360^5. \end{cases}$$

Simplifying gives

$$\begin{cases} 500 = 5 + 81 + 324 + 90 \\ 5 \cdot 81 \cdot 324 \cdot 90 \cdot (5 + 81 + 324 + 90) = 90^5. \end{cases}$$

Example 3.3. With $s = 5, 6$, we list here 10 small solutions to (3.15) $(a_1 = 1)$. Unfortunately, no counterexamples to Euler's conjecture have been found for $n = 6$.

a_1	a_2	a_3	a_4	b	n	a_1	a_2	a_3	a_4	a_5	b	n
1	2	12	12	6	27	1	1	2	2	2	2	8
1	4	4	18	6	27	1	6	6	6	8	6	27
1	4	20	25	10	50	1	1	9	9	16	6	36
1	3	32	36	12	72	1	2	3	12	18	6	36
1	4	12	64	12	81	1	9	12	18	24	12	64
1	3	8	96	12	108	1	4	16	24	27	12	72
1	27	36	64	24	128	1	6	9	24	32	12	72
1	1	18	108	12	128	1	4	8	32	36	12	81
1	25	54	100	30	180	1	4	12	16	48	12	81
1	4	27	256	24	288	1	2	9	36	48	12	96

Moving on, we ask for which n (3.15) admits infinitely many positive rational solutions; here we provide a partial answer to this question, omitting the proof.

Theorem 3.5. *For $s \geq 4$ and fixed positive integer n, if (3.15) has a solution $(a'_1, a'_2, \ldots, a'_{s-1} \; b^1)$ in positive rational numbers, and the rank of the elliptic curve*

$$Y^2 = X^3 - 27u'^3 v' \left(u'v'^3 - 24 \right) X + 54u'^4 (u'^2 v'^6 - 36u'v'^3 + 216)$$

is positive, then (3.15) has infinitely many positive rational solutions.

Remark 3.1. *Theorem* 3.2 *actually provides a new proof of Fermat's Last Theorem in the special case of exponent* $n = 3$, *one which is quite a bit simpler than Euler's proof, which relied on Fermat's method of infinite descent, as well as the properties of unique factorization and various congruences in the ring of integers of the imaginary quadratic field* $Q(\sqrt{-3})$.

For $s \geq 4$, we have the following question.

Question 3.4. For prime numbers $s \geq 4$, does the equation,

$$\begin{cases} n = a_1 + a_2 \\ a_1 a_2 (a_1 + a_2) = b^s \end{cases}$$

ever admit a pair of relatively prime positive integer solutions a_1 and a_2?

Obviously, this is a generalization of Fermat's Last Theorem, in the sense that the latter can be derived from a negative answer to this question.

On the other hand, consider the equation

$$\begin{cases} n = a_1 + a_2 \\ a_1^{\frac{p-1}{2}} a_2^{\frac{p-1}{2}} (a_1 + a_2) = b^p \end{cases}$$

where p is any odd prime number; the second of these equations becomes

$$u^{\frac{p-1}{2}}(u+1) = v^p. \tag{3.20}$$

Conjecture 3.2. *Equation* (3.20) *admits no solutions in rational numbers other than the trivial solutions* $(0,0)$ *and* $(-1,0)$; *this, too, easily implies Fermat's Last Theorem if it can be proved.*

For $s \geq 4$ and a given positive integer n, it is clear that (3.15) has at most finitely many positive integer solutions. The following question still remains.

Question 3.5. For $s \geq 4$, what is the smallest positive integer n such that (3.15) has a solution in positive integers.

When $s = 4, 5, 6$, it is easy to identify the corresponding smallest positive n, which are $n = 18, 27, 8$ respectively. In fact,

$$\begin{cases} 18 = 1 + 8 + 9 \\ 1 \cdot 8 \cdot 9 \cdot (1 + 8 + 9) = 6^4 \end{cases}$$

$$\begin{cases} 27 = 1 + 2 + 12 + 12 \\ 1 \cdot 2 \cdot 12 \cdot 12 \cdot (1 + 2 + 12 + 12) = 6^5 \end{cases}$$

$$\begin{cases} 8 = 1 + 1 + 2 + 2 + 2 \\ 1 \cdot 1 \cdot 2 \cdot 2 \cdot 2 \cdot (1 + 1 + 2 + 2 + 2) = 2^6. \end{cases}$$

Now, for $s \geq 4$ and given positive integer n, let $N(n, s)$ denote the number of solutions to (3.15) in positive integers.

Question 3.6. Is there a formula for the determination of $N(n, s)$ for given $s \geq 4$ and positive integer n?

According to Theorem 3.5, (3.15) has infinitely many positive rational solutions for given $s \geq 4$ and positive integer n under certain conditions, including a condition on the rank of an elliptic curve, which is by no means easily verified. We raise the following question.

Question 3.7. Assume $s \geq 2$; for given positive rational numbers u, v does the system

$$\begin{cases} u = b_1 \cdots b_s \\ v = b_1 + \cdots + b_s \end{cases}$$

of Diophantine equations admit infinitely many solutions in positive rational numbers?

When $s = 2$, it is easy to see that with

$$u = \frac{v^2 - w^2}{4}, \quad v > w$$

we get the solution

$$b_1 = \frac{v + w}{2}, \quad b_2 = \frac{v - w}{2}.$$

In 2014, Maciej Ulas proved (see Ulas, 2014) that for $s \geq 4$ and any non-zero real numbers A, B, the system

$$\begin{cases} A = x_1 x_2 \cdots x_s \\ B = x_1 + x_2 + \cdots + x_s \end{cases}$$

of Diophantine equations has infinitely many solutions, with $s - 3$ free parameters. In his proof, $x_3 = -4At^2 x_1$, from which it follows that if $A > 0$, then x_1 and x_3 cannot be simultaneously positive; for this reason, we do not obtain an answer to our question from this result.

We turn next to a different form of Diophantine equation. In 2005, Amarnath Murty proposed (see Murty, 2005) the following conjecture.

Conjecture 3.3. *For any $n > 3$, there exist positive integers a, b satisfying*

$$\begin{cases} n = a + b \\ ab - 1 = p \end{cases}$$

where p is a prime number; for any $n > 1$, when $n \neq 6, 30, 54$, there exist positive integers a, b such that

$$\begin{cases} n = a + b \\ ab + 1 = p. \end{cases}$$

where p is a prime number.

In 2012, the author proposed the following similar conjectures.

Conjecture 3.4. *For any $n > 14$, there exist positive integers a, b, c such that*

$$\begin{cases} n = a + b + c \\ abc + 1 = p^2. \end{cases}$$

Conjecture 3.5. *For any $n > 8$, there exist positive integers a, b, c, d such that*

$$\begin{cases} n = a + b + c + d \\ abcd = p^3 - p. \end{cases}$$

Remarks 3.2 We add the following explanation for Conjecture 3.2: from

$$\begin{cases} n = a + b \\ ab \pm 1 = p \end{cases}$$

and Vieta's formulas for polynomial equations, a and b two solutions for $x^2 - nx + p \mp 1 = 0$; that is:

$$\frac{n \pm \sqrt{n^2 - 4(p \mp 1)}}{2}.$$

Therefore, we also have a positive integer m such that

$$n^2 - 4(p \mp 1) = m^2.$$

If $n = 2k$, then $m = 2s$,

$$k^2 \pm 1 = p + s^2. \qquad (3.21)$$

If $n = 2k + 1$ then $m = 2s + 1$,

$$k^2 + k \pm 1 = p + s(s + 1). \qquad (3.22)$$

Here (3.21) is a consequence of the following famous conjecture.

Hardy-Littlewood Conjecture (1923). *Every sufficiently large even number is either a square or the sum of a prime and a square.*

For any odd prime number p, we call the rational number $\frac{p-1}{2}$ a half-prime. The twin prime conjecture is equivalent to the statement that there are infinitely many adjacent halfprimes. The Goldbach conjecture says that every positive integer larger than 1 is the sum of two half-primes.

Similarly, (3.22) is a consequence of the following statement: every positive integer larger than 1 is the sum of a half-prime and a triangular number. This statement is also consistent with a conjecture proposed (see Sun, 2009) by Sun Zhiwei, that every odd number greater than 3 can be expressed as $p + x(x + 1)$, where p is prime, and x a positive integer.

Chapter 4

Representation of Integers as Sums of Squares

My knowledge and my success were all acquired through diligent study.

—*Carl Friedrich Gauss*

4.1 Representations of Integers of Sums of Squares

We have seen in the first section of Chapter 2 that the study of square sum representations of integers dates to as early as the period of Fermat, who discovered and proved that odd prime numbers of the form $4n + 1$ can be written as a sum of the squares of two integers; moreover, this representation is unique up to reordering the summands. On the other hand, odd primes of the form $4n + 3$ cannot be expressed as a sum of two squares. The latter observation is obvious, since $x^2 + y^2 \equiv 0, 1$ or $2 \pmod 4$ for any integers x, y; specifically, $x^2 + y^2 \not\equiv 3 \pmod 4$. As for the former claim, there are various proofs. First, we prove it using Fermat's method of infinite descent; we then give another proof using the pigeonhole principle and various congruence properties (see Cai, 2021); a third proof invokes the relevant properties of the ring of Gaussian integers (see Hardy–Wright, 1979); finally, we present a constructive proof.

Proof 1. For prime $p \equiv \pmod 4$, it is well known that the Legendre symbol $\left(\frac{-1}{p}\right) = 1$, or in other words that there exists an integer t

such that

$$t^2 + 1 \equiv 0 \,(\mathrm{mod}\, p), \quad 0 < t < \frac{p}{2}.$$

Then

$$0 < 1 + t^2 < 1 + \frac{p^2}{4} < p^2.$$

We have

$$1 + t^2 = mp, \ 0 < m < p.$$

Let m_0 be the smallest positive integer m satisfying.

$$x^2 + y^2 = mp, \ p \nmid x, \ p \nmid y. \tag{4.1}$$

If $m_0 = 1$, we are done. Supposing instead that $1 < m_0 < p$, it is easy to see that m_0 cannot divide both x and y, since otherwise it would also divide p, which is impossible. Therefore there exist integers c, d such that

$$x_1 = x - cm_0, \quad y_1 = y - dm_0$$

satisfying

$$|x_1| \le \frac{m_0}{2}, \quad |y_1| \le \frac{m_0}{2}, \quad x_1^2 + y_1^2 > 0.$$

Therefore,

$$x_1^2 + y_1^2 \equiv x^2 + y^2 \equiv 0 \,(\mathrm{mod}\, m_0)$$

so we have

$$x_1^2 + y_1^2 = m_1 m_0 \tag{4.2}$$

where $0 < m_1 < m_0$. Multiplying (4.2) and (4.1) together, put $m = m_0$, and use the Fibonacci identity

$$(a^2 + b^2)(c^2 + d^2) = (ac \pm bd)^2 + (ad \mp bc)^2$$

to get

$$m_0^2 m_1 p = (x^2 + y^2)(x_1^2 + y_1^2) = (xx_1 + yy_1)^2 + (xy_1 - x_1 y)^2.$$

This Fibonacci, also known as Leonardo Pisano (1175–1250) is the Italian mathematician famous for proposing the "rabbit problem" known today as the Fibonacci sequence. Now note that

$$xx_1 + yy_1 = x(x - cm_0) + y(y - dm_0) = m_0 X$$
$$xy_1 - x_1 y = x(y - dm_0) - y(x - cm_0) = m_0 Y$$

where $X = p - cx - dy, Y = cy - dx$. Therefore,

$$m_1 p = X^2 + Y^2, \quad (0 < m_1 < m_0),$$

contradicting the minimality hypothesis on m_0. This completes the first proof.

The method of infinite descent cannot be used to prove the uniqueness of the representation; the following two methods prove both existence and uniqueness. □

Proof 2. As above, for primes $p \equiv 1 \,(\mathrm{mod}\,4)$, the Legendre symbol $\left(\frac{-1}{p}\right) = 1$, so that there exist positive integers t with

$$t^2 + 1 \equiv 0 \,(\mathrm{mod}\,p), (t, p) = 1. \tag{4.3}$$

Consider the values $tx - y$, with x, y drawn from the numbers $0, 1, \ldots, [\sqrt{p}]$; in total there are $([\sqrt{p}]+1)^2 > p$ such values. According to the pigeonhole principle, there must exist two distinct sets $\{x_1, y_1\}$ and $\{x_2, y_2\}$ satisfying

$$t^{x_1} - y_1 \equiv tx_2 - y_2 (\mathrm{mod}\,p).$$

Recalling that $(t, p) = 1$, $(t, p) = 1$, it is easy to see that $x_1 \neq x_2$, $y_1 \neq y_2$. We can assume without loss of generality that $y_1 > y_2$, $y = y_1 - y_2$, $x = \pm(x_1 - x_2)$, so that

$$ty \equiv x (\mathrm{mod}\,p)$$

where $0 < x, y < \sqrt{p}$.

Since $(y, p) = 1$, we can find an integer y^{-1} satisfying $yy^{-1} \equiv 1 (\mathrm{mod}\,p)$, so $t \equiv \pm xy^{-1} (\mathrm{mod}\,p)$. Substituting this into (4.3), we obtain $x^2 + y^2 \equiv 0 \,(\mathrm{mod}\,p)$; and since $0 < x^2 + y^2 < 2p$, we must have $x^2 + y^2 = p$.

Next, we prove uniqueness of the representation. Suppose

$$p = x^2 + y^2 = a^2 + b^2, \quad x > 0, \ y > 0, \ a > 0, \ b > 0.$$

Then

$$(ax - by)(ax + by) = a^2x^2 - b^2y^2 = a^2(x^2 + y^2) - y^2(a^2 + b^2)$$
$$\equiv 0 \,(\mathrm{mod}\ p)$$

from which it follows that $p|(ax - by)$ or $p|(ax + by)$. Invoking again the Fibonacci identity, we have

$$p^2 = (ax \mp by)^2 + (ay \pm bx)^2.$$

If $p|ax - by$, since $ay + bx > 0$, we must have $ax - by = 0$. Then from $(a, b) = (x, y) = 1$, we get $a = y$, $b = x$. Similarly if $p|ax + by$, then $ay - bx = 0$, $a = x$, $b = y$; this completes the proof of uniqueness. $\square$

Proof 3. As above, for $p \equiv 1 \,(\mathrm{mod}\ 4)$, the Legendre symbol $\left(\frac{-1}{p}\right) = 1$, so we can find an integer t such that

$$t^2 + 1 \equiv 0 \,(\mathrm{mod}\ p)$$

from which it follows that

$$p|(t + i)(t - i).$$

Now, if p remains prime in the ring of Gaussian integers, then it must divide one of $t + i$ or $t - i$, which is impossible, since

$$\frac{t}{p} \pm \frac{i}{p}$$

is not a Gaussian integer. It follows that p is not prime in the ring of Gaussian integers. If a Gaussian integer $\pi|p$, then also the conjugate ρ of π divides p, and we can write $p = \pi\rho$, where $\pi = a + bi$, $\rho = a - bi$ are Gaussian primes. Therefore,

$$p = a^2 + b^2.$$

That is, p can be written as a sum of two squares. We prove uniqueness via this argument as follows.

The factors of p in the ring of Gaussian integers are

$$\pm\pi, \ \pm\pi i, \ \pm\rho, \ \pm\rho i \qquad (4.4)$$

to which eight factors there correspond exactly the only representations of p as a sum of two squares:

$$p = (\pm a)^2 + (\pm b)^2 = (\pm b)^2 + (\pm a)^2.$$

If $p = c^2 + d^2$, then $c + di | p$ so $c + di$ is among the eight numbers described in (4.4) and therefore differs from the other representations only in sign and the order of the summands; this proves uniqueness.

The last proof, given by the Soviet mathematician I. M. Vinogradov (see Vinogradov 1), is constructive. □

Proof 4. Suppose p is prime and congruent to 1 modulo 4, and take a, b, respectively, any quadratic residue and quadratic nonresidue modulo p. Then we will prove that

$$p = \left(\frac{S(a)}{2}\right)^2 + \left(\frac{S(b)}{2}\right)^2$$

where

$$S(k) = \sum_{x=0}^{p-1} \left(\frac{x\left(x^2 + k\right)}{p}\right).$$

We show first that $S(k)$ is even for every integer k with $(k, p) = 1$. Note that

$$S(k) = \sum_{x=0}^{p-1} \left(\frac{x\left(x^2 + k\right)}{p}\right) = \sum_{x=0}^{2m} \left(\frac{x\left(x^2 + k\right)}{p}\right)$$
$$+ \sum_{x=2m+1}^{4m} \left(\frac{x\left(x^2 + k\right)}{p}\right)$$

where $m = (p - 1)/4$. Make the change of variables $x = p - y$ in the second term on the right to get

$$\sum_{x=2m+1}^{4m} \left(\frac{x\left(x^2 + k\right)}{p}\right) = \sum_{y=1}^{2m} \left(\frac{(p - y)\left((p - y)^2 + k\right)}{p}\right)$$
$$= \sum_{y=1}^{2m} \left(\frac{y\left(y^2 + k\right)}{p}\right).$$

So

$$S(k) = 2\sum_{x=0}^{2m} \left(\frac{x\left(x^2 + k\right)}{p}\right)$$

is even.

Next, we prove that $S(kt^2) = \left(\frac{t}{p}\right) S(k)$. If $p|t$, it is easy to see that both sides of the equation are zero, so there is nothing to prove. Suppose $p \nmid t$, and recall that when x passes through all values in a reduced residue system modulo p, so does xt; we have

$$S(kt^2) = \sum_{x=0}^{p-1} \left(\frac{x\left(x^2 + kt^2\right)}{p}\right) = \sum_{x=0}^{p-1} \left(\frac{xt\left((xt)^2 + kt^2\right)}{p}\right)$$

$$= \left(\frac{t}{p}\right) \sum_{x=0}^{p-1} \left(\frac{x\left(x^2 + k\right)}{p}\right) = \left(\frac{t}{p}\right) S(k),$$

as claimed.

Let $q = \frac{p-1}{2}$. From the above formula, we obtain

$$q^{S^2(a)} + q^{S^2(b)} = \sum_{t=1}^{q} S^2\left(at^2\right) + \sum_{t=1}^{q} S^2(bt^2).$$

Noting that $a \cdot 1^2, \ldots, 1^2, \ldots, a \cdot q^2$ and $b \cdot 1^2, \ldots, 1^2, \ldots, b \cdot q^2$ pass through a reduced system modulo p,

$$q^{S^2(a)} + q^{S^2(b)} = \sum_{k=1}^{p-1} S^2(k) = \sum_{x=1}^{p-1}\sum_{y=1}^{p-1}\sum_{k=0}^{p-1} \left(\frac{xy\left(x^2 + k\right)\left(y^2 + k\right)}{p}\right).$$

Continuing, we need the following identity:

$$\sum_{x=1}^{p-1} \left(\frac{x^2 + cx}{p}\right) = \sum_{x=1}^{p-1} \left(\frac{x^2 + x}{p}\right) = -1.$$

where $p \geq 3$, $p \nmid c$. This is because

$$\sum_{x=1}^{p-1} \left(\frac{x^2 + cx}{p} \right) = \sum_{x=1}^{p-1} \left(\frac{(cx)^2 + c(cx)}{p} \right) = \sum_{x=1}^{p-1} \left(\frac{x^2 + x}{p} \right)$$

$$= \sum_{x=1}^{p-1} \left(\frac{y^2 \left(x^2 + x \right)}{p} \right) = \sum_{x=1}^{p-1} \left(\frac{1 + y}{p} \right)$$

$$= \sum_{x=1}^{p-1} \left(\frac{1 + x}{p} \right) = -1$$

where y is the multiplicative inverse of x modulo p, that is, $yx \equiv 1 \pmod{p}$; in particular, as x passes through a reduced system modulo p, so does y.

It follows from this that when $y^2 \equiv x^2 \pmod{p}$,

$$\sum_{k=1}^{p-1} \left(\frac{xy \left(x^2 + k \right) \left(y^2 + k \right)}{p} \right) = \left(\frac{xy}{p} \right) \sum_{k=1, p \nmid x^2 + k}^{p-1} 1 = (p - 2) \left(\frac{xy}{p} \right).$$

Also when $y^2 \equiv x^2 \pmod{p}$, using the above summation formula with $c = y^2 - x^2$, we have

$$\sum_{k=1}^{p-1} \left(\frac{xy \left(x^2 + k \right) \left(y^2 + k \right)}{p} \right) = \left(\frac{xy}{p} \right) \sum_{j=x^2+1}^{x^2+p-1} \left(\frac{j \left(j + y^2 - x^2 \right)}{p} \right)$$

$$= -2 \left(\frac{xy}{p} \right).$$

Combining the above results,

$$q^{S^2(a)} + q^{S^2(b)} = p \sum_{x,y=1, x^2 \equiv y^2 (p)}^{p-1} \left(\frac{xy}{p} \right) + -2 \sum_{x,y}^{p-1} \left(\frac{xy}{p} \right)$$

$$= 2p(p - 1) + 0 = 4pq$$

that is,

$$p = \left(\frac{S(a)}{2}\right)^2 + \left(\frac{S(b)}{2}\right)^2.$$

The sum of two square integers discussed above is a special case of the quadratic form $ax^2 + bxy + cy^2$, where a, b, c are integers and $d = b^2 - 4ac$ is called the its discriminant. Research into quadratic forms primarily concerns whether and how integers can be presented as quadratic forms. We move on to consider when arbitrary positive integers m, not necessarily prime, can be written as a sum of two squares. □

Theorem 4.1. *Suppose $n > 0$ is $n = n_1^2 n_2$ where n_2 is squarefree (that is, has no square factors). Then n can be written as a sum of two square integers if and only if n_2 has no prime factors of the form $4m + 3$.*

Proof of Theorem 4.1. One direction is easy: if n_2 has no prime factors of the form $4m + 3$, then each of its prime factors can be written as a sum of two squares by the preceding argument (and $2 = 1^2 + 1^2$); then, using the Fibonacci identity again, it is clear that this condition is sufficient.

To show that also it is necessary, suppose $p = 4m + 3$ divides n_2, say $p^r \| n$ where $r \geq 1$. It is clear from the hypotheses of Theorem 4.1 that r is odd. Now suppose for the sake of contradiction that $n = x^2 + y^2$. Set $(x, y) = d$, and write

$$x = dx_1, \quad y = dy_1, \quad (x_1, y_1) = 1, \quad n = d^2(x_1^2 + y_1^2).$$

Since r is an odd prime, $p | x_1^2 + y_1^2$ and $(p, x_1) = 1$. Otherwise, $p | x_1, p | y_1$, contradicting the fact that $(x_1, y_1) = 1$. Therefore,

$$x_1^2 + y_1^2 \equiv 0 \,(\text{mod}\, p), \quad (p, x_1) = 1.$$

From basic modular arithmetic, we know that there exists an integer x_1' such that $x_1 x_1' \equiv 1 \,(\text{mod}\, p)$, and therefore

$$\left(y_1 x_1'\right)^2 \equiv -1 \,(\text{mod}\, p)$$

or $\left(\frac{-1}{p}\right) = 1$. On other hand, the known properties of the Legendre symbol give $\left(\frac{-1}{p}\right) = (-1)^{2m+1} = -1$, contradiction. This completes the proof of Theorem 4.1.

In particular, an odd prime number p admits a presentation as a sum of two square integers if and only if it is of the form $4m + 1$. Furthermore, suppose $n = 2^\alpha \prod_{p^r\|n} p^r \prod_{q^s\|n} q^s$, where p, q are primes of the form $4m + 1, 4m + 3$ respectively. Introducing the methods of complex variables, it is possible to prove the following theorem.

Theorem 4.2. *Let $r(n)$ be the number of ways to write n as a sum of two square integers. Then*

$$r(n) = 4 \prod_{\substack{p^r\|n \\ p\equiv 1(\bmod\ 4)}} (1+r) \prod_{\substack{q^s\|n \\ q\equiv 3(\bmod\ 4)}} \frac{1 + (-1)^2}{2}$$

or

$$r(n) = 4\delta(n),$$

where

$$\delta(n) = \sum_{d|n} k(d),$$

$$k(n) = \begin{cases} 0, & \text{if } n \text{ is even} \\ (-1)^{(n-1)/2} & \text{if } n \text{ is odd} \end{cases}$$

is a completely multiplicative arithmetic function.

As an application of Theorem 4.2, we get the following result. The system of equations

$$\begin{cases} n = a + b \\ ab = x^2 \end{cases}$$

has a solution in positive integers if and only if n is even or n has prime factors of the form $4m + 1$.

4.2 The Four-Squares Theorem

We turn now to the statements and proofs of two famous theorems.

Theorem 4.3 (Lagrange). *Every positive integer can be expressed as a sum of the squares of four integers.*

Theorem 4.4 (Gauss). *Any positive integer except for those of the form $4^k(8n + 7)$ can be expressed as a sum of the squares of three integers.*

From Theorem 4.4, it follows that integers of the form $8n + 3$ can be expressed as a sum of three squares, from which Theorem 4.3 can be deduced. We consider first prime numbers p. If $p = 2$, the result is obvious; if $p \equiv 1 \pmod 4$, then the theorem in the previous section shows that it can be represented as a sum of two squares; and if $p \equiv 3 \pmod 4$, then one of p or $p - 4$ is congruent to 3 modulo 8, and therefore can be expressed as a sum of three square integers. Since 4 itself is also a square, it follows that every prime can be expressed a sum of four square integers.

As for arbitrary integers, we need only to note that if $m = x_1^2 + x_2^2 + x_3^2 + x_4^2$, $n = y_1^2 + y_2^2 + y_3^2 + y_4^2$, then

$$mn = (x_1y_1 + x_2y_2 + x_3y_3 + x_4y_4)^2 + (x_1y_2 - x_2y_1 + x_3y_4 - x_4y_3)^2$$
$$+ (x_1y_3 - x_3y_1 + x_4y_2 - x_2y_4)^2 + (x_1y_4 - x_4y_1 + x_2y_3 - x_3y_2)^2$$

completing the proof of Theorem 4.3.

Lagrange (Fig. 4.1), Euler, Gauss (Fig. 4.2), these three were perfect mathematicians, who could see on the one hand the magnificent properties of the natural numbers, and on the other could point with their fingers to the stars in the vastness of space (Fig. 4.3).

We first prove Theorem 4.2 (see Hua 1, Chapter 6, Section 7), for which we require the following lemmas.

Lemma 4.1 (see Cai, 2021, Corollary to Theorem 4.7). *For any positive integer $n > 1$, the number of solutions $v(n)$ to the congruence*

$$x^2 \equiv -1 \pmod n \tag{4.5}$$

is given by

$$v(n) = \begin{cases} 0, & 4|n \\ \prod_{p|n}(1 + k(p)), & 4 \nmid n \end{cases} \tag{4.6}$$

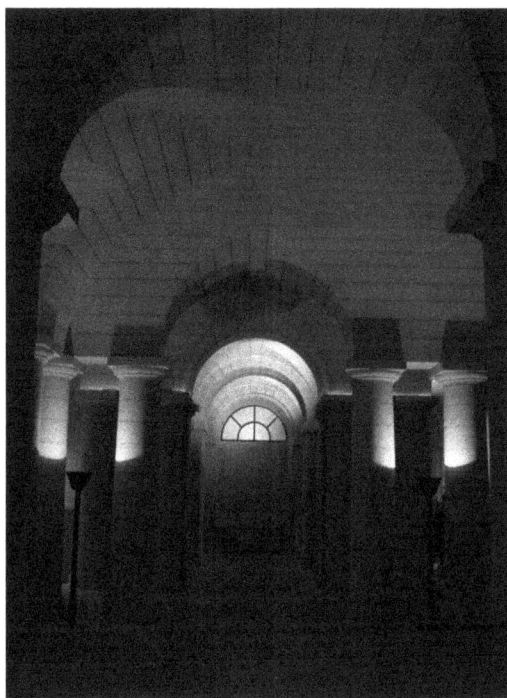

Figure 4.1. The Pantheon in Paris, where Lagrange is buried. Photograph by the author.

Figure 4.2. Portrait of Gauss.

Figure 4.3.   Statue of a Goose Girl, photograph by the author in Göttingen.

Lemma 4.2. *Let $n > 1$; then corresponding to any solution l of*

$$l^2 \equiv -1 \,(\mathrm{mod}\, n)$$

there exists a pair of integers (xy) satisfying

$$x^2 + y^2 = n, \quad y \equiv lx \,(\mathrm{mod}\, n).$$

Proof of Theorem 4.2. From Lemmas 4.1 and 4.2, we see that the number of solutions to

$$x^2 + y^2 = n, \quad (x, y) = 1$$

is given by $4v(n)$. More generally, considering the solutions to $x^2 + y^2 = n$ with $(x, y) = d$, we see that the number of solutions to

$$\left(\frac{x}{d}\right)^2 + \left(\frac{y}{d}\right)^2 = \frac{n}{d^2}$$

is given by $4v\left(\frac{n}{d^2}\right)$, so

$$r(n) = 4\sum_{d^2|n} V\left(\frac{n}{d^2}\right) = 4\sum_{d|n} v\left(\frac{n}{d}\right)\lambda(n)$$

where

$$\lambda(n) = \begin{cases} 1, & \text{if } d \text{ is a perfect square} \\ 0 & \text{otherwise.} \end{cases}$$

Since $v(n)$ and $\lambda(n)$ are both multiplicative functions, so is $\frac{r(n)}{4}$. Then since $\delta(n)$ is also multiplicative, we need only to prove that

$$\frac{r(n)}{4} = \delta(n)$$

when $n = \boldsymbol{p^m}$ to complete the proof of Theorem 4.2.

If $2|m$, then

$$\frac{r(p^m)}{4} = v\left(p^m\right) + v\left(p^{m-2}\right) + \cdots + v\left(p^2\right) + v(1)$$

$$= \begin{cases} 0 + \cdots + 0 + 1 = 1, & \text{if } p = 2 \\ 0 + \cdots + 0 + 1 = 1, & \text{if } p \equiv 1 \pmod 4 \\ 2 + \cdots + 2 + 1 = 2\cdot\frac{m}{2} + 1 = m + 1, & \text{if } p \equiv 3 \pmod 4 \end{cases}$$

If $2 \nmid m$, then we instead have

$$\frac{r(p^m)}{4} = \begin{cases} 1, & \text{if } p = 2 \\ 0, & \text{if } p \equiv 1 \pmod 4 \\ m + 1, & \text{if } p \equiv 3 \pmod 4 \end{cases}$$

On the other hand,

$$\delta(p^m) = \delta(p^m) = 1 + k(p) + \cdots + k\left(p^m\right)$$

$$= \begin{cases} 1 + 0 \cdots + 0 = 1, & \text{if } p = 2 \\ 1 - 1 + \cdots + 1 = 1, & \text{if } p \equiv 3 \pmod 4 \quad \text{and } 2|m \\ 1 - 1 + \cdots - 1 = 0, & \text{if } p \equiv 3 \pmod 4 \quad \text{and } 2 \nmid m \\ 1 + 1 + \cdots + 1 = m + 1, & \text{if } p \equiv 1 \pmod 4. \end{cases}$$

as required. This completes the proof of Theorem 4.2.

To prove Theorem 4.3, we need only the following lemma, which was originally derived by both Lagrange and Euler (see Erickson-Vazzana, 2006).

Lemma 4.3. *For any odd prime p, there exist integers x, y, and k such that*

$$x^2 + y^2 + 1 = kp \qquad (4.7)$$

where $0 < k < p$.

Proof. We need to prove that there are integers x, y satisfying

$$x^2 + y^2 + 1 \equiv 0 \pmod{p}$$

with $x^2 + y^2 + 1 < p^2$. We consider separately two cases.

If $p \equiv 1 \pmod 4$, then $\left(\frac{-1}{p}\right) = 1$, so the congruence

$$x^2 \equiv -1 \pmod{p}$$

has solutions. We can assume that $0 < x < p/2$; then

$$x^2 + 1 < \frac{p^2}{4} + 1 < p^2.$$

Since p divides $x^2 + 1$, we get a solution to (4.7), with $y = 0$.

If $p \equiv 3 \pmod 4$, let a be the smallest quadratic nonresidue modulo p. From $\left(\frac{-1}{p}\right) = -1$, we can see that $\left(\frac{-a}{p}\right) = 1$, so there exists x such that

$$x^2 \equiv -a \pmod{p}$$

with $0 < x < p/2$. By the minimality condition on a, $a - 1$ is a quadratic residue modulo p, so there also exists y such that

$$y^2 \equiv a - 1 \pmod{p}$$

with $0 < y < p/2$. Here $x^2 + y^2 \equiv -1 \pmod p$ and

$$x^2 + y^2 + 1 < (p/2)^2 + (p/2)^2 + 1 < p^2$$

completing the proof of Lemma 4.3.   □

Proof of Theorem 4.3. From Lemma 4.3, we see that there exist integers $x, y, z,$ and w such that

$$x^2 + y^2 + z^2 + w^2 = kp$$

with $0 < k < p$. We choose x, y, z, w such that k is as small as possible. If $k = 1$, then we are done. Otherwise, suppose $k > 1$, and we will endeavor to find a contradiction; in fact, we again employ Fermat's method of infinite descent.

Observe first that k cannot be an even number; otherwise there must be an even number of odd numbers among x, y, z, w; then we can assume that $x + y, x - y, z + w$ and $z - w$ are all even, and we get

$$\left(\frac{x+y}{2}\right)^2 + \left(\frac{x-y}{2}\right)^2 + \left(\frac{z+w}{2}\right)^2 + \left(\frac{z-w}{2}\right)^2 = \left(\frac{k}{2}\right)p,$$

contradicting the minimality of k.

Therefore, let k be odd. There exist integers a, b, c, d such that

$$a \equiv x \pmod{k}$$
$$b \equiv y \pmod{k}$$
$$c \equiv z \pmod{k}$$
$$d \equiv w \pmod{k}$$

and $0 \le |a|, |b|, |c|, |d| < k/2$. We have

$$a^2 + b^2 + c^2 + d^2 \equiv x^2 + y^2 + z^2 + w^2 \equiv 0 \pmod{k}$$

so there exists a non-negative integer m such that $a^2 + b^2 + c^2 + d^2 = mk$. It is easy to see that $m \neq 0$, since otherwise $a = b = c = d = 0$, implying that $k^2 | x^2 + y^2 + z^2 + w^2$, contradiction. On the other hand,

$$a^2 + b^2 + c^2 + d^2 < \left(\frac{k}{2}\right)^2 + \left(\frac{k}{2}\right)^2 + \left(\frac{k}{2}\right)^2 + \left(\frac{k}{2}\right)^2 = k^2$$

so $m < k$.

From Euler's four-square identity, we know that the product of $(a^2+b^2+c^2+d^2)$ and $(x^2+y^2+z^2+w^2)$ is also a sum of four squares. In fact, with

$$X = xa + yb + zc + wd$$
$$Y = xb - ya + zd - wc$$
$$Z = xc - za + wb - yd$$
$$W = xd - wa + yc - zd$$

we get

$$X^2 + Y^2 + Z^2 + W^2 = \left(a^2 + b^2 + c^2 + d^2\right)\left(x^2 + y^2 + z^2 + w^2\right)$$
$$= (mk)(kp). \tag{4.8}$$

Note that

$$X \equiv x^2 + y^2 + z^2 + w^2 \equiv 0 \pmod{k},$$
$$Y \equiv xy - yx + zw - wz \equiv 0 \pmod{k},$$
$$Z \equiv xz - zx + wy - yw \equiv 0 \pmod{k},$$
$$W \equiv xw - wx + yz - zy \equiv 0 \pmod{k},$$

or in other words k divides each of X, Y, Z, and W, so from (4.8) we obtain

$$\left(\frac{X}{k}\right)^2 + \left(\frac{Y}{k}\right)^2 + \left(\frac{Z}{k}\right)^2 + \left(\frac{W}{k}\right)^2 = mp$$

where $0 < m < k$, contradiction.

Since 2 can be written as a sum of four squares, we have proved that every prime can be written as a sum of four squares, and so Theorem 4.3 follows by Euler's four-square identity.

4.3   The Three-Squares Theorem

In this section, we prove Gauss's three-squares theorem (Theorem 4.4); the proof here is due to Ankeny (1957), and uses

Dirichlet's theorem on arithmetic progressions, as well as the following theorem due to Minkowski on the geometry of numbers (see Hardy–Wright, 1979, Theorem 446).

Lemma 4.4. *Every convex set in n-dimensional space with volume greater than 2^n contains a point with all integer coordinates, not all of which are zero.*

Proof of Theorem 4.4. We assume first that m is a squarefree positive integer congruent to 3 modulo 8; write $m = p_1 p_2 \cdots p_r$, where the p_j are prime numbers. Let q be a prime number satisfying the following conditions:

$$\left(\frac{-2q}{p_j}\right) = 1, \quad (1 \le j \le r) \tag{4.9}$$

$$q \equiv 1 \pmod 4 \tag{4.10}$$

where $\left(\frac{a}{b}\right)$ is the Jacobi symbol. It is easy to conclude that such a prime q must exist from Dirichlet's theorem and considering a reduced system of residues modulo $4m$.

By (4.9) and (4.10),

$$1 = \prod_{j=1}^{r}\left(\frac{-2q}{p_j}\right) = \prod_{j=1}^{r}\left(\frac{-2}{p_j}\right)\left(\frac{q}{p_j}\right)$$

$$= \left(\frac{-2}{m}\right)\prod_{j=1}^{r}\left(\frac{p_j}{q}\right) = \left(\frac{-2}{m}\right)\left(\frac{m}{q}\right)$$

$$= \left(\frac{-2}{m}\right)\left(\frac{-m}{q}\right) = \left(\frac{-m}{q}\right). \tag{4.11}$$

Here, we have made use of (4.10) and the fact that $m \equiv 3 \pmod 8$.

Since q is an odd prime, there exists an odd number b such that $b^2 \equiv -m \pmod q$, or

$$b^2 - qh_1 = -m. \tag{4.12}$$

Reducing both sides of (4.12) modulo 4, we get $1 - h_1 \equiv 1 \pmod 4$, so $h_1 = 4\hbar$ for some integer h. Then

$$b^2 - 4qh = -m \tag{4.13}$$

On the other hand, from (4.9) and the Chinese Remainder Theorem, we know that there exists an integer s such that $s^2 \equiv -2q(\bmod m)$, and therefore also an integer t such that

$$t^2 \equiv -1/2q(\bmod m). \tag{4.14}$$

We consider the geometric set

$$R^2 + S^2 + T^2 < 2m \tag{4.15}$$

where

$$R = 2tqx + tby + mz$$

$$S = (2q)^{\frac{1}{2}}x + \frac{b}{(2b)^{\frac{1}{2}}}y \tag{4.16}$$

$$T = \frac{m^{\frac{1}{2}}}{(2b)^{\frac{1}{2}}}y.$$

Clearly (4.15) represents a convex set with volume $(4/3)\pi(2m)^{3/2}$ and symmetric about the origin; moreover, the determinant of the linear transformation (4.16) is $m^{3/2}$. It follows that after substitution, (4.15) represents a convex set in the variables (x, y, z), symmetric about the origin in three-dimensional space, with a volume of $(2^{7/2}/3)\pi$. It is not hard to see that $(2^{7/2}/3)\pi > 8$.

So, it follows from Lemma 4.4 that there exists a points (x, y, z) with integer coordinates, not all zero, satisfying (4.15) and (4.16), which we write as (x_1, y_1, z_1); we write the corresponding values of (R, S, T) as (R_1, S_1, T_1). Then

$$R_1^2 + S_1^2 + T_1^2 = (2tqx_1 + tby_1 + mz_1)^2$$

$$+ \left((2q)^{\frac{1}{2}}x_1 + \frac{b}{(2q)^{\frac{1}{2}}}y_1 \right)^2 + \left(\frac{m^{\frac{1}{2}}}{(2q)^{\frac{1}{2}}}y_1 \right)^2 \tag{4.17}$$

$$\equiv t^2 (2qx_1 + by_1)^2 + \frac{1}{2q} (2qx_1 + by_1)^2$$

$$\equiv 0(\bmod m)$$

where we have made use of (4.14).

Furthermore,

$$R_1^2 + S_1^2 + T_1^2 = R_1^2 + \left((2q)^{\frac{1}{2}} x_1 + \frac{b}{(2q)^{\frac{1}{2}}} y_1 \right)^2 + \left(\frac{m^{\frac{1}{2}}}{(2q)^{\frac{1}{2}}} y_1 \right)^2$$

$$= R_1^2 + \frac{1}{2q} (2qx_1 + by_1)^2 + \frac{m}{2q} y_1^2$$

$$= R_1^2 + 2(qx_1^2 + bx_1 y_1 + hy_1^2). \tag{4.18}$$

Set

$$v = qx_1^2 + bx_1 y_1 + hy_1^2. \tag{4.19}$$

Note that R_1 is an integer; from (4.16), (4.17), (4.18) we find that $m|R_1^2 + 2v$, from (4.15) that $R_1^2 + 2v < 2m$, and from the nondegeneracy of (4.16), that (x_1, y_1, z_1) are not all zero; so

$$R_1^2 + 2v = m. \tag{4.20}$$

Suppose p is an odd prime divisor of v with odd multiplicity: $p^{2n+1} \| v$. If $p \nmid m$, then by (4.20),

$$\left(\frac{m}{p} \right) = 1. \tag{4.21}$$

Also, by (4.19) and (4.13),

$$4qv = (2qx_1 + by_1)^2 + (4qh - b^2)y_1^2 = (2qx_1 + by_1)^2 + my_1^2. \tag{4.22}$$

If $p|q$, then $(-m/p) = 1$ by (4.22).

If $p|q$, then likewise by (4.22), we have

$$p^{2n+1} \| e^2 + mf^2$$

so $\left(\frac{-m}{p} \right) = 1$. Therefore, in either case,

$$p \equiv 1 \pmod 4.$$

If $p|m$, then by (4.19), (4.20), and (4.13),

$$R_1^2 + \frac{1}{2q} \left((2qx_1 + by_1)^2 + my_1^2 \right) = m. \tag{4.23}$$

It follows that $p|R_1, p|(2qx_1 + by_1)$. Recalling that m is squarefree, dividing both sides of (4.25) by p gives

$$\frac{1}{2q}\frac{m}{p}y_1^2 \equiv \frac{m}{p}(\bmod\, p)m$$

or

$$y_1^2 \equiv 2q(\bmod\, p), \left(\frac{2q}{p}\right) = 1.$$

Comparing to (4.9), we find that $\left(\frac{-1}{p}\right) = 1$, so

$$p \equiv 1(\bmod\, 4).$$

To summarize, if v has an odd prime factor p with odd multiplicity, then necessarily p is congruent to 1 modulo 4; if not, then one of v or $2v$ is a perfect square, and it follows from (4.20) that m is a sum of three squares. We conclude that Theorem 4.4 holds in the case that $m \equiv 3(\bmod\, 8)$.

When $m \equiv 1, 2, 5$ or 6 $(\bmod\, 8)$, we can adjust the proof accordingly. Let q be such that, for any odd prime factor p_j of m, we have $\left(\frac{-2q}{p_j}\right) = 1$, and $q \equiv 1(\bmod\, 4)$. If m is even, let $m = 2^{m_1}$. Then

$$\left(\frac{-2}{q}\right) = (-1)(m_2 - 1)/2, t^2 \equiv -\frac{1}{q}(\bmod\, p_j),$$

where t is an odd number, $b^2 - qh = -m$.

Again let

$$R = tqx + tby + mz$$

$$S = q^{\frac{1}{2}}x + \frac{b}{q^{\frac{1}{2}}}y$$

$$T = \frac{m^{\frac{1}{2}}}{q^{\frac{1}{2}}}y.$$

Then the rest of the proof goes through as above; this completes the proof of Theorem 4.4.

If we extend $\delta(n)$ to $\Delta(n) = \#\{n = a + b, ab = c^2\}$, then $\Delta(n) \geq \delta(n)$. In 1996, Albert H. Beiler proved that if n is odd with $n = n_1 n_2$, where n_1, n_2 contain respectively all prime factors of n congruent to 1 modulo 4 and 3 modulo 4, then

$$\Delta(n) = \frac{1}{2} \left\{ \sum_{d|n_1} 2^{\omega(d)} - 1 \right\}$$

where $\omega(d)$ is the multiplicative function given by the number of distinct prime factors of d.

We (Zhong-Cai, 2017) have proved a more general results, and at the same time obtained that for even numbers $n = 2^\alpha n_1 n_2 (\alpha \geq 1, n_1, n_2$ as above), we have

$$\Delta(n) = \sum_{d|n_1} 2^{\omega(d)}.$$

As we have seen previously, primes congruent to 1 modulo 4 can be written as a sum of two squares, for example $13 = 2^2 + 3^2 = (2 + 3\sqrt{-1})(2 - 3\sqrt{-1})$. This corresponds to the fact that in the quadratic extension $Q(\sqrt{-1})$, such rational primes decompose as a product of two prime elements, while rational primes congruent to 3 modulo 4 remain prime. More generally, the problem of prime factorization in quadratic number fields, and the associated problem of the factorization of prime ideals in the ring of integers of a quadratic number field belong to class field theory, a branch of algebraic number theory that grew out of Fermat's proposition and the problem of quadratic reciprocity investigated by Gauss.

4.4 Products as Polygonal Numbers

In 1752, in a letter to Goldbach, Euler asked the following question: are there infinitely many prime numbers of the form $n^2 + 1$? This is a very difficult question, and nobody has been able to resolve it to this day. In 2016, we (Cai Tianxin and Chen Deyi), accidently discovered

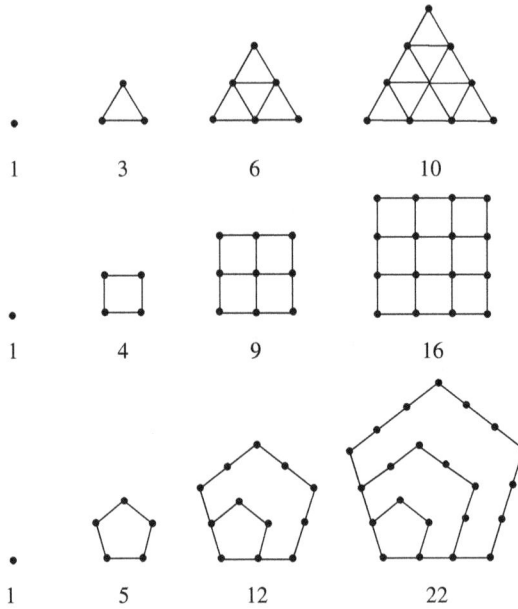

Figure 4.4. Polygonal numbers.

(see Zhong-Cai, 2017 or Cai 3) a necessary and sufficient condition for $n^2 + 1$ to be prime.

Theorem 4.5. *The additive and multiplicative equation*

$$\begin{cases} n = a + 2b \\ ab = \binom{c}{2} \end{cases} \tag{4.24}$$

has solutions in positive integers if and only if $n^2 + 1$ is composite.

Equivalently, $n^2 + 1$ is prime if and only if (4.24) has no solutions in positive integers.

Similarly, we also obtained the following theorem.

Theorem 4.6. *The additive and multiplicative equation*

$$\begin{cases} n = a + b \\ ab = \binom{c}{2} \end{cases} \tag{4.25}$$

has solutions in positive integers if and only if $2n^2 + 1$ is composite.

Here we prove only Theorem 4.6; the proof of Theorem 4.5 is similar.

Proof of Theorem 4.6. We note first that equation (4.25) has a solution if and only if $2n^2 + 1$ can also be represented as $2x^2 + y^2$, $y \geq 3$. One direction of this claim follows by setting $x = a - b, y = 2c - 1$; then $n = a + b$. For the other direction, since y is odd, n and x are either both even or both odd, and we can put $a = \frac{n+x}{2}, b = \frac{n-x}{2}, c = \frac{y+1}{2}$.

Let $T = 2n^2 + 1$. Then the number of solutions to (4.25) is one less than the numbers of ways to express T as $2x^2 + y^2$ where x and y are required to be positive integers. Keeping this last condition in mind, the mass formula (see Hua, 1982) gives

$$T = \frac{1}{2} \sum_{d | 2n^2+1} \left(\frac{-2}{d} \right)$$

where $\left(\frac{a}{b} \right)$ is the Kronecker symbol, which is the same as the Jacobi symbol when d is an odd number larger than 1. It follows that (4.25) has no solutions if and only if

$$\sum_{d | 2n^2+1} \left(\frac{-2}{d} \right) = 2.$$

If $d = p_1^{\alpha_1} \ldots p_k^{\alpha_k} | 2n^2 + 1$, then clearly each $p_i | 2n^2 + 1$ where the $p_i (1 \leq i \leq k)$ are odd primes. Therefore, since $p_i \nmid n$, we can find n' such that $nn' \equiv 1 \pmod{p_i}$, and from $2n^2 + 1 \equiv 0 \pmod{p_i}$ we find that

$$2 + (n')^2 \equiv 0 \pmod{p_i}, \quad \left(\frac{-2}{p_i} \right) = 1.$$

Then from the definition of the Jacobi (Kronecker) symbol, also $\left(\frac{-2}{d} \right) = 1$. So, (4.25) has no solution if and only if

$$\sum_{d | 2n^2+1} 1 = 2$$

in other words, if $2n^2 + 1$ is prime. This completes the proof of Theorem 4.5

Replacing the coefficient 2 in (4.24) with an odd prime, we do not always obtain a similar result. We do however have such a result in the cases $p = 3, 5, 11$, and 29; concretely, the equation

$$\begin{cases} n = a + pb \\ ab = \left(\frac{c}{2} \right) \end{cases}$$

has solutions in positive integers (with a possibly equal to 0) if and only if $2n^2 + p$ is composite.

In addition, the author makes the following two conjectures.

Conjecture 4.1. $\{primes\ p|2n^2 + p\ is\ prime\ for\ 0 \le n \le p-1\} = \{3, 5, 11, 29\}$.

Conjecture 4.2. $\{primes\ p|q \nmid 2n^2 + p\ for\ any\ prime\ q < p, n \in \mathbb{N}\} = \{3, 5, 11, 29\}$.

Moreover, we conjecture that the set of prime numbers of the form $2n^2 + 29$ is equal to the set of minimum factors of numbers of the form $2(29 + \Delta)^2 + 29$, and the set of prime factors of numbers of the form $2n^2 + 29$ is equal to the set of factors of numbers of the form $2(29i + \Delta/i)^2 + 29$ where Δ runs through all nonnegative triangular numbers and i is any positive integer such that Δ is a multiple of i.

In 2018, we (Cai Tianxin, *et al.*, preprint) further extended equation (4.25) to

$$\begin{cases} n = a + b \\ ab = tP(m, c) \end{cases} \tag{4.26}$$

where a, b, and t are positive integers, and

$$P(m, c) = \frac{c}{2}\{(m - 2)c - (m - 4)\}$$

is the cth m-gonal number. Obviously, when $m = 3$ or $m = 4, P(m, c)$ is a positive integer, and when $m > 4, P(m, c)$ is an integer. In this notation, we can write (4.24) as

$$\begin{cases} n = a + b \\ ab = c(c + 1). \end{cases}$$

Let $r_{m,t}(n)$ denote the number of solutions of equation (4.26). We investigate first the case where $r_{m,t}(n) = 0$, that is, when (4.26) has no solutions; in particular, the special case $r_{5,1}(n) = 0$, corresponding to the situation when

$$\begin{cases} n = a + b \\ ab = P(5, c) = \frac{1}{2}c(3c - 1) \end{cases}$$

has no solutions, and prove that this occurs if and only if $6n^2 + 1$ is prime.

More generally, we have the following theorem.

Theorem 4.7. *If* $2(m-2)n^2 + t(m-4)^2$ *is prime, then* (4.26) *has no solutions.*

In order to prove Theorem 4.7, we will need the following lemma (see Nagell, 1951 Theorem 101).

Lemma 4.5. *For* c, d *nonnegative integers, a prime number* p *can be expressed in at most one way as*

$$cx^2 + dy^2$$

where x, y *are non-negative integers.*

We also introduce the notation $r'_{m,t}(n)$ for the equation

$$2(m-2)n^2 + t(m-4)^2 = 2(m-2)x^2 + ty^2 \qquad (4.27)$$

which we need to consider in the proof of Theorem 4.7.

Proof of Theorem 4.7. Suppose first that (a, b, c) is a solution to (4.26). Then $(x, y) = (|a-b|, |2(m-2)c - (m-4)c|)$ is a nonnegative integer solution to (4.27). This is because

$$2(m-2)(a-b)^2 + t(2(m-2)c - (m-4))^2$$
$$= 2(m-2)\big((a+b)^2 - 4ab\big) + t(2(m-2)c - (m-4))^2$$
$$= 2(m-2)(n^2 - 4tP(m,c)) + 4t(m-2)c((m-2)c - (m-4))$$
$$\quad + t(m-4))^2$$
$$= 2(m-2)n^2 - 8t(m-2)P(m,c) + 8t(m-2)P(m,c)$$
$$\quad + t(m-4))^2$$
$$= 2(m-2)\,n^2 + t(m-4))^2.$$

Next, if (4.26) has another solution (a', b', c') in integers satisfying

$$|a-b| = |a'-b'|, |2\,(m-2)c - (m-4)| = |2\,(m-2)\,c' - (m-4)|$$

then $a = a'$ or $b = a'$; so $r_{m,t}(n) \leq r'_{m,t}(n)$. And since a, b are greater than 1, in fact

$$r_{m,t}(n) \leq r'_{m,t}(n) - 1.$$

Recalling that $(x, y = (n, |m - 4|)$ is a solution to (4.27), then by Lemma 4.5, if $2(m - 2)\,n^2 + t(m - 4)^2$ is prime, then $r'_{m,t}(n) = 1$, $r_{m,t} r_{m,t}(n) = 0$, or in other words (4.26) has no solutions. This completes the proof of Theorem 4.7.

The relationship between $r_{m,t}(n)$ and $r'_{m,t}(n)$ is given by the following theorem.

Theorem 4.8. *We have*

$$r_{m,t}(n) = r'_{m,t}(n) - 1.$$

under any of the following three conditions

(1) $t = 1$, $m = 3$ *or* $p + 2$,
(2) $t = 2$, $m = 3$ *or* $2p + 2$,
(3) t *is an odd prime*, $t \neq m - 2, m = 3$ *or* $p + 2$, *where also* p *is an odd prime.*

Proof. Suppose (x, y) is an integer solution to (4.27), p is an odd prime. $\qquad \square$

Case 1: $t = 1$, $m = 3$ *or* $p + 2$.

If $n < x$, then $y < |m - 4|$, which is impossible for $m = 3$ or 5. When $m > 5, y < |m - 4|$, so $0 \leq y \leq m - 5$. If $y = 0$, we have

$$2pn^2 + (p - 2)^2 = 2px^2$$

from which we conclude that $p | (p - 2)^2$, contradiction. Therefore, we must have $1 \leq y \leq m - 5$. By (4.27),

$$2(m - 2)n^2 + (m - 4)^2 = 2(m - 2)x^2 + y^2. \qquad (4.28)$$

Therefore,

$$2(m - 2)(n + x)(n - x) = (y + m - 4)(y - m + 4).$$

so that $2p = 2m - 4$ divides one of $y + m - 4$ or $m - 4 - y$. Noting that

$$m - 3 \leq y + m - 4 \leq 2m - 9, 1 \leq m - 4 - y \leq m - 5$$

we have

$$m - 5 < 2m - 9 < 2m - 4$$

contradiction. Putting all of this together, we must have $n \geq x$.

When $n > x$, we verify directly that $((n + x)/2, (n - x)/2, c)$ is a solution in positive integers to (4.26), where

$$C = \frac{m - 4 + y}{2(m - 2)} \text{ or } \frac{m - 4 - y}{2(m - 2)}$$

is an integer, and a positive integer when $m > 3$. In fact, (4.28) implies that n and x have the same parity, so that $(n + x)/2$ and $(n - x)/2$ are positive integers. Then according to the previous discussion, when $n > x, 2p = 2m - 4$ divides $m - 4 + y$ or $m - 4 - y$, implying that one of $\frac{m-4+y}{2(m-2)}$ or $\frac{m-4-y}{2(m-2)}$ is an integer. So, we can consider an integer c satisfying $y = |2(m - 2)c - (m - 4)|$.

Now we can confirm that $((n + x)/2, (n - x)/2, c)$ is a solution to (4.26). Obviously $(n + x)/2 + (n - x)/2 = n$; on the other hand,

$$((n + x)/2) \cdot ((n - x)/2)) = \frac{y^2 - (m - 4)^2}{8(m - 2)}$$

$$= \frac{(2(m - 2)c - (m - 4))^2 - (m - 4)^2}{8(m - 2)}$$

$$= \frac{c}{2}((m - 2)c - (m - 4)) = P(m, c).$$

It follows that $(x, y) \to \left(\frac{n+x}{2}, \frac{n-x}{2}, c\right)$ is an injective map.

If $n = x$, then $(n, |m - 4|)$ is a solution in integers to (4.27), but $((n+x)/2, (n-x)/2, c)$ is not a solution in positive integers of (4.26), since in this case $(n - x)/2 = 0$.

Therefore, $r_{m,t}(n) \geq r'_{m,t}(n) - 1$; and from the proof of Theorem 4.7, we know that $r_{m,t}(n) = r'_{m,t}(n) - 1$.

Case 2: $t = 2, m = 3$ or $2p + 2$.

If $n < x$, then $y < |m - 4|$, impossible when $m = 3$. When $m > 5, y < |m-4|$, so $0 \leq y \leq m-5$. If $y = 0$, then $2p^{n^2+4(p-1)^2} = 2px^2$, from which it follows that

$$p | 4(p - 1)^2,$$

contradiction. So, we must have $1 \leq y \leq m - 5$.

From (4.27),

$$(m - 2)n^2 + (m - 4)^2 = (m - 2)x^2 + y^2. \tag{4.29}$$

Therefore,

$$(m - 2)(n + x)(n - x) = (y + m - 4)(y - m + 4)$$

$2p = 2m - 2$ must divide one of $y + m - 4$ or $m - 4 - y$. Noting that

$$m - 3 \le y + m - 4 \le 2m - 9, 1 \le m - 4 - y \le m - 5$$

we have

$$m - 5 < 2m - 9 < 2m - 4$$

contradiction. We conclude that $n \ge x$.

For $n > x$, we consider two cases. If $m = 3$, then $n^2 + 1 = x^2 + y^2$. Considering the congruence

$$n^2 + 1 \equiv x^2 + y^2 \pmod 4$$

we find that

$$n^2 \equiv x^2 \pmod 4 \text{ or } n^2 \equiv y^2 \pmod 4.$$

Assuming without loss of generality that $n^2 \equiv x^2 \pmod 4$, n and x must have the same parity, and y must be odd. Then $((n + x)/2, (n - x)/2, (y - 1)/2)$ is a solution in positive integers to (4.26):

$$((n + x)/2) + ((n - x)/2) = n$$

and

$$((n + x)/2) \cdot ((n - x)/2) = (y^2 - 1)/4$$
$$= ((2c + 1)^2 - 1)/4 = 2P(3, c).$$

When $m = 2p + 2$, we verify that $((n + x)/2, (n - x)/2, c)$ is a solution in positive integers to (4.26), where

$$c = \frac{m - 4 + y}{2(m - 2)} \text{ or } \frac{m - 4 - y}{2(m - 2)}$$

is an integer. From (4.28), we see that n and x have the same parity, so that $(n + x)/2$ and $(n - x)/2$ are positive integers. According to the previous discussion, $2p = 2m - 4$ divides $m - 4 + y$ or $m - 4 - y$, which implies that one of $\frac{m-4+y}{2(m-2)}$ or $\frac{m-4-y}{2(m-2)}$ is a positive integer. We consider an integer c satisfying

$$y = |2(m - 2)c - (m - 4)|.$$

We now confirm that $((n + x)/2, (n - x)/2, c)$ is a solution to (4.26). Obviously $((n + x)/2) + ((n - x)/2) = n$. On the other hand,

$$((n + x)/2) \cdot ((n - x)/2) = \frac{y^2 - (m - 4)^2}{4(m - 2)}$$

$$= \frac{(2(m - 2)c - (m - 4))^2 - (m - 4)^2}{4(m - 2)}$$

$$= c((m - 2)c - (m - 4)) = 2P(m, c).$$

It follows that $(x, y) \to (\frac{n+x}{2}, \frac{n-x}{2}, c)$ is an injective map.

If $n = x$, then $(n, |m - 4|)$ is a solution in integers to (4.27), but $((n+x)/2, (n-x)/2, c)$ is not a solution in positive integers to (4.26), because in this case $(n - x)/2 = 0$.

Therefore, $r_{m,t}(n) \geq r'_{m,t}(n) - 1$. Then from the proof of Theorem 4.7, we conclude that $r_{m,t}(n) = r'_{m,t}(n) - 1$.

Case 3: t is an odd prime, $t \neq m - 2, m = 3$ or $p + 2$.

This case is similar to the previous two; it is easy to see that $((n + x)/2, (n - x)/2, c)$ is a solution in positive integers to (4.26), where

$$c = \frac{m - 4 + y}{2(m - 2)} \text{ or } \frac{m - 4 - y}{2(m - 2)}$$

is an integer, positive when $m = 3$. Moreover, the case $n \leq x$ admits only a single solution given by $(m, |m - 4|)$, from which we conclude that $r_{m,t}(n) = r'_{m,t}(n) - 1$. This completes the proof of Theorem 4.8.

Using Theorem 4.8, and various related properties of quadratic forms such as their ranks, we can give various concrete values for $r_{m,t}(n)$; we omit the proofs.

Theorem 4.9. *Let* $d(n)$ *indicate the number of divisors of* n. *Then*

$$r_{3,1}(n) = [\{d\left(2n^2 + 1\right) - 1\}/2]$$

$$r_{5,1}(n) = [\{d\left(6n^2 + 1\right) - 1\}/2]$$

$$r_{7,1}(n) = [\{d_{\{3\}}\left(10n^2 + 9\right) - 1\}/2]$$

$$r_{13,1}(n) = [\{d_{\{3\}}\left(22n^2 + 81\right) - 1\}/2]$$

$$r_{31,1}(n) = [\{d_{\{3\}} (58n^2 + 729) - 1\}/2]$$

$$r_{3,2}(n) = d(n^2 + 1)/2 - 1$$

$$r_{8,2}(n) = [\{d_{[2]} (3n^2 + 8) - 1\}/2]$$

$$r_{12,2}(n) = [\{d_{[2]} (5n^2 + 32) - 1\}/2]Y$$

$$r_{24,2}(n) = [\{d_{[2,5]} (11n^2 + 200) - 1\}/2]$$

$$r_{60,2}(n) = [\{d_{[2,7]} (29n^2 + 1568) - 1\}/2]$$

$$r_{3,3}(n) = [\{d_{[3]} (2n^2 + 3) - 1\}/2]$$

$$r_{3,5}(n) = [\{d_{[5]} (2n^2 + 5) - 1\}/2]$$

$$r_{3,11}(n) = [\{d_{[11]} (2n^2 + 11) - 1\}/2]$$

$$r_{3,29}(n) = [\{d_{[29]} (2n^2 + 29) - 1\}/2]$$

where [x] is the largest integer not exceeding x, $d_A(n) = \prod_{p \notin A}$ $(1 + \text{ord}_p n)$; in particular, $d_\emptyset(n) = d(n)$ is the number of divisors of n.

Corollary 4.1. *Let $\mathbb{P}$ be the set of all prime numbers. Then*

$$r_{3,1}(n) = 0 \Leftrightarrow 2n^2 + 1 \in \mathbb{P},$$

$$r_{5,1}(n) = 0 \Leftrightarrow 6n^2 + 1 \in \mathbb{P},$$

$$r_{7,1}(n) = 0 \Leftrightarrow 10n^2 + 9 \in \mathbb{P} \cup 9\mathbb{P},$$

$$r_{13,1}(n) = 0 \Leftrightarrow 22n^2 + 81 \in \mathbb{P} \cup 9\mathbb{P} \cup 81\mathbb{P},$$

$$r_{31,1}(n) = 0 \Leftrightarrow 58n^2 + 729 \in \mathbb{P} \cup 9\mathbb{P} \cup 81\mathbb{P} \cup 729\mathbb{P},$$

$$r_{3,2}(n) = 0 \Leftrightarrow n^2 + 1 \in \mathbb{P},$$

$$r_{8,2}(n) = 0 \Leftrightarrow 3n^2 + 8 \in \mathbb{P} \cup 4\mathbb{P} \cup 8\mathbb{P},$$

$$r_{12,2}(n) = 0 \Leftrightarrow 5n^2 + 32 \in \mathbb{P} \cup 4\mathbb{P} \cup 16\mathbb{P} \cup 32\mathbb{P},$$

$$r_{24,2}(n) = 0 \Leftrightarrow 11n^2 + 200 \in \mathbb{P} \cup 4\mathbb{P} \cup 25\mathbb{P} \cup 100\mathbb{P} \cup 200\mathbb{P},$$

$$r_{60,2}(n) = 0 \Leftrightarrow 29n^2 + 1568 \in \mathbb{P} \cup 4\mathbb{P} \cup 16\mathbb{P} \cup 32\mathbb{P} \cup 49\mathbb{P} \cup 196$$
$$\times \mathbb{P} \cup 784\mathbb{P} \cup 1568\mathbb{P},$$

$$r_{3,3}(n) = 0 \Leftrightarrow 2n^2 + 3 \in \mathbb{P} \cup 3\mathbb{P},$$

$$r_{3,5}(n) = 0 \Leftrightarrow 2n^2 + 5 \in \mathbb{P} \cup 5\mathbb{P},$$
$$r_{3,11}(n) = 0 \Leftrightarrow 2n^2 + 11 \in \mathbb{P} \cup 11\mathbb{P},$$
$$r_{3,29}(n) = 0 \Leftrightarrow 2n^2 + 29 \in \mathbb{P} \cup 29\mathbb{P}.$$

In 1978, the Polish-American mathematician Henryk Iwaniec, 1947) proved (see Iwaniec, 1978) that there are infinitely many integers n such that the irreducible quadratic polynomial $an^2 + bn + c$ has at most two prime factors, where $a > 0, c$ is an odd number. From this, we obtain the following corollary.

Corollary 4.2. *If $(m,t) = (3,1), (5,1)$ or $(7,1)$, then there exist infinitely many n such that $r_{m,t}(n) \leq 1$.*

Finally, we can consider systems similar to (4.25) but with more independent variables, for example,

$$\begin{cases} n = a + b + c \\ abc = \binom{d}{3}. \end{cases}$$

We conjecture that when $n \neq 1, 2, 4, 7$, or 11, this additive and multiplicative equation admits solutions in positive integers.

Chapter 5

Figurate Primes and F-Perfect Numbers

Mathematics is the art of ingenuity.

—*Paul Halmos*

5.1 Definition of Figurate Primes

In 1742, shortly after his transfer from the St. Petersburg Academy of Sciences to the Berlin Academy of Sciences, the Swiss mathematician Euler remarked in a letter to Goldbach that any even number greater than 4 can be expressed as a sum of 2 odd prime numbers. Goldbach had previously written to Euler with his own observation that *every odd number greater than or equal to 9 can be expressed as a sum of three odd prime numbers.* The latter problem is known as the ternary Goldbach conjecture. It is easy to see that the ternary Goldbach conjecture is a direct consequence of Euler's conjecture, and later generations have referred to them collectively as the Goldbach conjecture, perhaps because there are already so many theorems and formulas named after Euler in the history of mathematics. In any case, people later learned from the rediscovered lost writings of the Frenchman Descartes that as early as the 17th century, that geometer who also loved number theory had discovered Goldbach's conjecture among the mysteries of the natural numbers.

Although any primary student can verify that $6 = 3 + 3, 8 = 3 + 5, 10 = 3 + 7 = 5 + 5 \cdots$, nevertheless nobody has been able to prove or disprove the Goldbach conjecture. The best result was obtained by the Chinese mathematician Chen Jingrun (1933–1996), who proved in 1966 that:

> "every sufficiently large even number can be written as the sum of a prime number and an odd number with no more than two prime factors."

Chen Jingrun used a new weighted sieve method, a variation on the ancient sieve of Eratosthenes. The Soviet mathematician Ivan Vinogradov (1891–1983) had already proved as early as 1937 that every sufficiently large odd number can be written as a sum of three prime numbers, using the circle method invented by Hardy and Littlewood. In May 2013, the Peruvian mathematician Harald Helfgott (1977–), working at the École Normale Supérieure in Paris, France, published two papers (see Helfgott, 2014) announcing a complete proof of the ternary Goldbach conjecture.

It should be pointed out that many results related to the Goldbach conjecture (for example, concerning the number of representations) can be extended respectively to the twin prime conjecture, which states that there are infinitely many pairs of prime numbers separated by a difference of 2, such as $(3,5)$ and $(5,7)$. Historians of mathematics have been unable to work out when, where, and by whom the twin prime conjecture was first formulated. We do know for certain that the French mathematician Alphonse de Polignac (1826–1863) proposed the generalization of the twin prime conjecture that is now known by his name.

De Polignac Conjecture. For any natural number k, there exist infinitely many pairs of prime numbers p and q such that $p - q = 2k$.

When $k = 1$, this is precisely the twin prime conjecture.

In 2004, the Chinese-Australian mathematician Terence Tao (1975–) and the British mathematician Ben Green (1977–) used ergodic theory, a topic in analysis, and Ramsey theory, from combinatorics, to prove the following theorem.

Green–Tao Theorem. *There exist infinitely many prime arithmetic progressions of arbitrary length.*

Here, length refers to the number of elements in the arithmetic progression; for example, 3, 5, 7 is a prime arithmetic progression of length 3, with common difference 2, 109, 219, 329, 439, 549 is a prime arithmetic progression of length 5 with common difference 110.

In 2007, the Polish mathematician Jarosław Wróblewski found a prime arithmetic progression of length 24

$$468395662504823 + 45872132836530n(0 \leq n \leq 23)$$

in 2008 and 2010, others obtained prime arithmetic progressions of lengths 25 and 26, respectively.

The Green–Tao theorem is very strong. Prior to its proof, people could not even show that there existed infinitely many prime arithmetic progressions of length 3. In 2006, Tao won the Fields Medal on the basis of this result and other work.

On the other hand, as early as 1940, Erdős proved that there exists a constant $c < 1$ such that there exist infinitely many pairs of prime numbers p, p' satisfying $p' - p < c \ln p$. In 2005, the American Daniel Goldston, Hungarian János Pintz, and the Turkish mathematician Cem Yıldırım collaborated to prove that c can be any small positive number. This work was the basis of the proof by Chinese mathematician Zhang Yitang (1955–) in spring of 2013 of the following result.

Theorem (Zhang Yitang). *There are infinitely many pairs (p, q) of prime numbers separated by a difference of no more than 7×10^7.*

This result signifies an important step towards a proof of the twin prime conjecture, and Zhang Yitang was awarded numerous honors for its achievement. Subsequently, the Polymath8 project inaugurated by Terence Tao obtained successive improvements in decreasing the constant bound in the theorem. In particular, a young Oxford University postdoctoral fellow named James Maynard (1987–) introduced an original method to bring the constant down to 600, and later, other mathematicians used his method to reduce it further to 246; this was in March of 2014. Since then there have however been no further improvements. This result reminds me of the famous classical Chinese mathematical masterpiece, *Nine Chapters on the Mathematical Art*, which contains exactly 246 problems. At the 2020 International Congress of Mathematicians, Maynard was awarded the Fields Medal for this and other work.

For a long time I have maintained three reservations about the Goldbach conjecture, or, in other words, I view it as having three shortcomings. First, prime numbers are the atoms of integer multiplication, and not well suited to constructions of integer addition; secondly, separately representing even and odd numbers respectively as sums of two and three primes is not consistent with mathematical beauty; and finally, as n increases, there appear more and more ways to express it as a sum of odd primes, leading to wasteful redundancy.

To this end, after many exploratory investigations and calculations, the author defined (see Cai *et al.*, 2015) in Spring of 2013 the figurate primes, given by

$$\binom{p^i}{j},$$

where p is a prime number, and i and j are any positive integers. The set of figurate primes includes 1 and all prime numbers and prime powers, among which even numbers occur relatively rarely but nevertheless infinitely often. It can be shown that the figurate primes exhibit the characteristics of both the prime numbers and the figurate numbers (a topic belonging to the ancient Pythagorean school), and that there are as many figurate primes as there are prime numbers in the sense of natural density: the number of figurate primes not exceeding a positive number x is $x/\log x$.

We have the following conjecture (verified up to 10^7).

Conjecture 5.1. *Every positive integer larger than 1 can be written as a sum of two figurate primes.*

Moving on, if we define proper figurate primes to be figurate primes that are not also prime numbers, we easily estimate the order of proper figurate primes to be $C\sqrt{x}/\log x$, where $C = 2+2\sqrt{2}$, with odd and even numbers each accounting for about half of the total. We summarize the data in the following table.

Range	Number of primes	Number of figurate primes	Number of proper figurate primes
≤ 100	25	47	22
≤ 1000	168	226	58
≤ 10000	1229	1355	126
≤ 100000	9592	9866	274
≤ 1000000	79498	79096	598
Order	$x/\log x$	$x/\log x$	$C\sqrt{x}/\log x$

We make the following conjecture (verified up to 10^7).

Conjecture 5.2. *Any integer greater than 5 can be represented as a sum of a prime number and a proper figurate prime.*

At the same time, we also present a more sophisticated conjecture than the twin prime conjecture,

Conjecture 5.3. *For any positive integer k, there are infinitely many pairs of figurate primes separated by difference k. In particular, there exist infinitely many pairs of adjacent figurate primes ($k = 1$).*

If both p and $\frac{p^2-p+2}{2}$ are prime, then $\binom{p}{2}$ and $\binom{p}{2} + 1$ comprise a pair of adjacent figurate primes, and conversely. The prime numbers smaller than 100 satisfying this requirement are $2, 5, 13, 17, 41, 61, 89, 97$; the 1000th such prime is 116797.

If we exclude 1 and the trivial symmetry of the binomial coefficients, we have also the following conjecture.

Conjecture 5.4. *The figurate primes consist of all different numbers.*

Conjecture 5.4 is the one we expect to be proved first, as it is less difficult than any of the previous conjectures. Although we cannot yet give a complete proof, we have the following result.

Theorem 5.1. *If $\binom{p^\alpha}{i} = \binom{q^\beta}{j}$ non-trivially, where α and β are positive integers, and $0 < i < p^\alpha, 0 < j < q^\beta$, then it must be the case that either $p \mid i$, or $q \mid j$.*

This theorem was proposed by the graduate student Chen Xiaohang in the fall of 2015 during our number theory seminar. To prove it, we need the following lemma, which was first established in 1852 by the German mathematician Ernst Kummer, and rediscovered in 1878 by the French mathematician Édouard Lucas (1842–1891).

Lemma 5.1. *If a and b are positive integers, and m is the highest power of prime p dividing $\binom{a+b}{a}$, then m is given by the number of carries when a and b are expanded in base p and added to one another.*

See Ribenboim (1995) for the proof; from the contents of the proof, we also obtain another lemma.

Lemma 5.2. *If $p^m \| \binom{n}{k}, 1 \leq k \leq n$, then $p^m \leq n$.*

Proof of Theorem 5.1. Suppose for the sake of contradiction that $p \nmid i, q \nmid j$ and for any $l \leq i_0 < i$, assume $p^v \| i_0, 0 \leq v < \alpha$. Then $p^v \| p^\alpha - i_0$. From this, we can see that the 2nd through ith terms in the numerator and the 1st through $(i-1)$th terms in the denominator of the fraction $\binom{p^a}{i} = \frac{p^a(p^a-1)\ldots(p^a-(i-1))}{1\ldots(i-1)i}$ contain respectively the same powers of p, which therefore cancel out, so that $p^\alpha \| \binom{p^a}{i}$. By the same logic, $q^\beta \| \binom{q^\beta}{j}$. By Lemma 5.2, $p^\alpha \leq q^\beta$ and $q^\beta \leq p^\alpha$, so $p^\alpha = q^\beta$, $i = j$ or $p^a - j$, contradicting the hypothesis. This completes the proof of Theorem 5.1.

In 1900, at the International Congress of Mathematicians at Sorbonne University in Paris, David Hilbert spoke the following words, which will live on forever in the history of mathematics, at the end of his presentation of his Problem 8:

> "After an exhaustive discussion of Riemann's prime number formula, perhaps we may sometime be in a position to attempt the rigorous solution of Goldbach's problem, viz., whether every integer is expressible as the sum of two positive prime numbers; and further to attack the well-known question, whether there are an infinite number of pairs of prime numbers with the difference 2, or even the more general problem, whether the linear diophantine equation $ax + by + c = 0$ (with given integral coefficients each prime to the others) is always solvable in prime numbers x and y."[1]

Up to this point, we have seen no specific question or conjecture related to the linear diophantine equation mentioned by Hilbert; having introduced the concept of figurate primes, we tried to give this diophantine equation a definite meaning, including the Goldbach conjecture and the twin primes conjecture, which correspond respectively to the cases $a = b = 1$ with n even and $a = 1, b = -1, c = -2$. After carrying out a numerical survey by computer, we have the following conjecture, the second half of which can be derived from Schinzel's (Fig. 5.1) hypothesis and the properties of the diophantine equation.

---

[1] Bulletin (New Series) of the American Mathematical Society Volume 37, Number 4, pp. 407–436. S 0273-0979(00)00881-8; Article electronically published on June 26, 2000.

Figure 5.1.   The author and Schinzel on a train in Kyushu, Japan.

Conjecture 5.5. *Let a and b be positive integers with $(a, b) = 1$. Then for every integer $n > (a-1)(b-1) + 1$, the equation*

$$ax + by = n$$

always has a solution (x, y) in figurate primes; if also $n \equiv a + b$ (mod 2), then the equation

$$ax - by = n$$

has infinitely many prime solutions (x, y).

5.2   Generalization of Pillai's Conjecture

In 1912, the American-born British mathematician Louis Joel Mordell (1888–1972) studied the equation

$$y^2 = x^3 + k, \tag{5.1}$$

where k is an integer, and proposed as a conjecture that for any integer k, (5.1) has at most finitely many integer solutions. In 1922, he proposed (see Mordell, 1922) the following more general conjecture.

Figure 5.2.   Portrait of Pillai.

The Mordell Conjecture. If C is a curve of genus greater than 1 over the field of rational numbers, then there are at most finitely many rational points on C.

In 1983, the German mathematician Gerd Faltings proved this conjecture, which is now known as Faltings's theorem; as a corollary, this proved that the Fermat equation has at most finitely many solutions. In 1986, Faltings was awarded the Fields Medal at the International Congress of Mathematicians held at the University of California, Berkely.

Earlier, in 1931, the Indian mathematician S.S. Pillai (Fig. 5.2) had proposed (see Pillai, 1931) the conjecture that was later named after him.

Pillai's Conjecture. If a and b are positive integers greater than 1, then the equation

$$x^a - y^b = k \tag{5.2}$$

has at most finitely many solutions for any fixed integer k.

If x and y are required to be prime, and a and b in equation (5.2) are allowed to be 1, then we have the following special cases: with x and odd prime, $y = 2, k = 1$, we get the Fermat primes; with $x = 2, b = 1, k = 1$, we get Mersenne primes; when $a = b = 1, k = 2$, we have the twin prime conjecture; when $k = 1$, we have Catalan's conjecture (now Mihăilescu's theorem), which we discuss in Section 7.3.

In fact, there is a more general form of Pillai's conjecture: if A, B, C are fixed positive integers, $(m, n) \neq (2, 2)$, then the diophantine equation

$$Ax^m - By^n = C$$

has at most finitely many solutions in integers.

Like the Beal conjecture and the Fermat–Catalan conjecture, it is easy to see that Pillai's conjecture holds under the assumption that the abc conjecture is true. Sadly, Pillai died in a plane crash while traveling from Kalkota en route to the Institute for Advanced Study in Princeton and the International Congress of Mathematicians in Boston.

In light of the definition of figurate primes, we consider the following equation:

$$\binom{p^a}{i} - \binom{q^b}{j} = k \tag{5.3}$$

where p and q are prime numbers, a, b, i, j, k are positive integers. Using elementary methods and techniques in number theory, we can prove the following theorem.

Theorem 5.2. *Fix* $k = 1, j = 1$. *When* $i = 2$, *(5.3) has exactly four solutions:* $(p, q, a, b) = (2, 5, 2, 1), (3, 2, 1, 1), (2, 3, 3, 3)$, *and* $(5, 3, 1, 2)$; *when* $i = 3$, *(5.3) has exactly three solutions:* $(p, q, a, b) = (2, 3, 2, 1), (3, 83, 2, 1), (5, 3, 1, 2)$; *when* $i = 4$, *(5.3) has exactly two solutions:* $(p, q, a, b) = (5, 2, 1, 2), (3, 5, 2, 3)$.

With $i = b = 1, j = 2$, then if a is even, (5.3) has the unique solution $(p, q) = (2, 3)$; if $a = 1$, then it seems that (5.3) has infinitely many solutions. For example, there are infinitely many (p, q) satisfying

$$p^a - 1 = \binom{q}{2}.$$

However, it is difficult to express all solutions concretely. Then 10 smallest solutions are $(pq) = (2, 2), (11, 5), (79, 13), (137, 17), (821, 41), (1831, 61), (3917, 89), (4657, 97), (5051, 101)$, and

(6329, 113). For odd $a > 1$, we conjecture that (5.3) has no solutions, which is easy to prove in the case that $a = 3$.

Similarly, we can also consider the diophantine equation

$$p^a - 1 = \binom{q^b}{3}$$

with solutions $(p, q, a, b) = (5, 2, 1, 2), (2, 3, 1, 1), (11, 5, 1, 1)$.

When $i = j \geq 2$, it is easy to verify that (5.3) has no solutions.

Using the theory of elliptic curves and the Magma software package, we get the following two theorems.

Theorem 5.3. *If* $(i, j) = (2, 3)$, *then* (5.3) *has the unique solution* $(p, q, a, b) = (3, 7, 2, 1)$; *if* $(i, j) = (3, 2)$, *then* (5.3) *has exactly two solutions* $(p, q, a, b) = (2, 3, 2, 1), (3, 7, 2, 1)$; *if* $(i, j) = (2, 4)$, *then again* (5.3) *has two solutions* $(p, q, a, b) = (2, 5, 2, 1), (3, 7, 2, 1)$; *finally, if* $(i, j) = (4, 2)$, *then* (5.3) *has no solutions.*

Theorem 5.4. *Fix* $k = 2$. *If* $(i, j) = (2, 3)$, *then* (5.3) *has exactly two solutions* $(p, q, a, b) = (2, 2, 2, 2), (3, 3, 1, 1)$; *if* $(i, j) = (3, 2)$, *then* (5.3) *has no solutions; if* $(i, j) = (2, 4)$, *then* (5.3) *has the unique solution* $(p, q, a, b) = (3, 2, 1, 2)$; *finally, if* $(i, j) = (4, 2)$, *then* (5.3) *has the unique solution* $(p, q, a, b) = (5, 3, 1, 1)$.

Proof of Theorem 5.2. Fix $k = 1$. If $(i, j) = (2, 1)$, then (5.3) is equivalent to

$$(p^a + 1)(p^a - 2) = 2q^b.$$

It is easy to see that $d = (p^a + 1, \; p^a - 2) = 1$ or 3.

When $d = 1$, $p = 2$, we have

$$2^a - 2 = 2, \quad 2^a + 1 = q^b$$

or

$$2^a - 2 = 2q^b, \quad 2^a + 1 = 1$$

from which we obtain $(p, q, a, b) = (2, 5, 2, 1)$.

When $d = 1, p > 2$, we have

$$p^a - 2 = q^b, \quad p^a + 1 = 2$$

or

$$p^a - 2 = 1, \quad p^a + 1 = 2q^b$$

from which we obtain $(p, q, a, b) = (3, 2, 1, 1)$.
   When $d = 3, q = 3$, or $p = 2$, we have

$$2^a - 2 = 6, \quad 2^a + 1 = 3^{b-1}$$

or

$$2^a - 2 = 2 \cdot 3^{b-1}, \quad 2^a + 1 = 3$$

from which we get $(p, q, a, b) = (2, 3, 3, 3)$.
   Finally, if $p > 2$, we have

$$p^a - 2 = 3, p^a + 1 = 2 \cdot 3^{b-1}$$

or

$$2^a - 2 = 3^{b-1}, 2^a + 1 = 6, p^a - 2 = 3^{b-1}, p^a + 1 = 6$$

from which we get $(p, q, a, b) = (5, 3, 1, 2)$.
   The cases $(i, j) = (3, 1)$ or $(4, 1)$ are handled similarly, completing the proof of Theorem 5.2.
   In the following arguments, we use the notation $(X, \pm Y) = (10, -11; 189)$ to indicate that $(X, \pm Y) = (10, 189)$ or $(-11, 189)$.

Proof of Theorem 5.3. Fix $k = 1$. For convenience, if $(i, j) = (2, 3)$, then set $p^a = y, q^b = x$ in (5.3), and if $(i, j) = (3, 2)$, instead set $p^a = x, q^b = y$.
   Let

$$x = \frac{X + 12}{12}, \quad y = \frac{Y + 36}{72}$$

with inverse transformation

$$X = 12x - 12, \quad Y = 36(2y - 1).$$

We obtain, respectively,

$$Y^2 = X^3 - 144X + 11664, \quad Y^2 = X^3 - 144X - 3024.$$

Magma can easily find the integer points of these two elliptic curves. In the first case, we get the integer point $(72, 612)$, from which we get the solution $(p, q, a, b) = (3, 7, 2, 1)$ to (5.3). In the second case, we get the two integer points $(36, 180)$ and $(84, 756)$, from which we obtain two solutions, $(p, q, a, b) = (2, 3, 2, 1)$ and $(p, q, a, b) = (3, 7, 2, 1)$ to (5.3).

If $(i,\ j) = (2, 4)$, set $p^a = y$, $q^b = x$ in (5.3). Let

$$x = \frac{X + 3}{6}, \quad y = Y + 2$$

from which we get

$$Y^2 = 3X^4 + 6X^3 - 3Y^2 - 6X + 81.$$

We again use Magma to find the integer points on this elliptic curve:

$$(X, \pm Y) = (10, -11; 189), (-1, -2; 9), (1, 0; 9), (-4, 3; 21),$$
$$(-6, 5; 51), (-92, 91, 14499).$$

We find that when $(X, Y) = (3, 21)$ or $(5, 51)$ there are two solutions to (5.3) given by $(p, q, a, b) = (2, 5, 2, 1), (3, 7, 2, 1)$.

If $(i, j) = (4, 2)$, set $p^a = x, q^b = y$ in (5.3). Then

$$x = \frac{x + 3}{6}, \quad y = Y + 2,$$

so

$$Y^2 = 3X^4 + 6X^3 - 3Y^2 - 6X + 81,$$
$$Y^2 = 3X^4 + 6X^3 - 3X^2 - 6X - 63.$$

Using Magma, we find the integer points

$$(X, \pm Y) = (-3, 2; 3).$$

It is easy to see that this does not produce any solution to (5.3). This completes the proof of Theorem 5.3.

Proof of Theorem 5.4. Set $k = 2$. For convenience, if $(i, j) = (2, 3)$, set $p^a = y, q^b = x$ in (5.3), and if $(i, j) = (3, 2)$, set $p^a = x, q^b = y$.
Then

$$x = \frac{X + 12}{12}, \quad y = \frac{Y + 36}{72}$$

with inverse transformation

$$X = 12x - 12, \quad Y = 36(2y - 1).$$

From this, we obtain, respectively,

$$Y^2 = X^3 - 144X + 22032, \quad Y^2 = X^2 - 144X - 19440.$$

We use Magma to find the integer points of these two elliptic curves; in the first case, there are two integer points given by $(36, 252)$ and $(24, 180)$, from which we get the solutions $(p, q, a, b) = (2, 2, 2, 2)$ and $(3, 3, 1, 1)$ to (5.3); in the second case, there are no integer points, and correspondingly no solutions to (5.3).

If $(i, j) = (2, 4)$, set $p^a = y, q^b = x$. in (5.3). Then

$$x = X, \quad y = \frac{Y + 3}{6}$$

from which we get

$$Y^2 = 3X^4 - 18X^3 + 33X^2 - 18X + 153.$$

Magma gives the integer points

$$(X, \pm Y) = (-1, 4; 15).$$

We find that when $(X, Y) = (4, 15)$, there is a unique solution $(p, q, a, b) = (3, 2, 1, 2)$ to (5.3).

If $(i, j) = (4, 2)$, set $p^a = x, \ q^b = y$, in (5.3). Then

$$x = X, \quad y = \frac{Y + 8}{6}$$

from which we get

$$Y^2 = 3X^4 - 18X^3 + 33X^2 - 18X - 135. \qquad (5.4)$$

Using Magma, we find the integer points to be

$$(X, \pm Y) = (-2, 5; 15)$$

and it follows that when $(X, Y) = (5, 15)$, there is a unique solution $(p, q, a, b) = (5, 3, 1, 1)$ to (5.3). This completes the proof of Theorem 5.4.

5.3   The F-Perfect Number Problem

In 2000, the Clay Institute in New York announced a list of *Millenium Problems*, comprising a total of seven mathematical problems, and offered a reward of one million US dollars for the solution of each of them. So far, only the Poincaré conjecture among them has been resolved, by the Russian mathematician Grigori Perelman (1966–), who refused to accept the substantial prize money. Also in the year 2000, the Italian mathematician Piergiorgio Odifreddi (1950–), recipient of the Galileo Prize and the Peano Prize, published a book entitled *The Mathematical Century: The 30 Greatest Problems of the Last 100 Years*, expounding upon thirty problems that sparked major breakthroughs in the course of the 20th century. In it, he presented four unresolved problems, three of which overlapped with the Millenium Problems of the Clay Institute. *The perfect number problem*, first on his list, was not, however, among these.

A perfect number is a positive integer such that the sum of its proper divisors (that is, not including itself) is equal to itself; symbolically, a percet number is a positive integer satisfying the equation

$$\sum_{d|n,d<n} d = n. \tag{5.5}$$

It is believed that the ancient Greek mathematician Pythagoras had investigated the perfect numbers in the sixth century BCE and knew already that 6 and 28 are perfect numbers, since

$$6 = 1 + 2 + 3,$$
$$28 = 1 + 2 + 4 + 7 + 14.$$

Pythagoras wrote that six symbolizes a perfect marriage, as well as health and beauty, because its parts are complete and the sum of its parts is equal to itself.

Working around the third century BCE, Euclid provided the above definition of perfect numbers and gave a sufficient condition for an even number to be perfect: if both p and $2^p - 1$ are prime numbers, then

$$2^{p-1}\left(2^p - 1\right) \tag{5.6}$$

is a perfect number.

This condition and its proof appear as Proposition 36 in Book IX of Euclid's *Elements*; Proposition 20 of the same book proves that there are infinitely many prime numbers.

In the *Book of Genesis*, the first volume of the Old Testament of the Bible, it is written that God created the world in six days, the seventh day taken for rest. About this, the ancient Roman philosopher Saint Augustine (354–430) wrote that "Six is a number perfect in itself, and not because God created the world in six days; rather, the contrary is true: God created the world in six days because this number is perfect."

The third and fourth perfect numbers are 496 and 8128, respectively. These two were mentioned in the *Introduction to Arithmetic* by Nicomachus of Gerasa (ca. 60–120 CE), a member of the New Pythagorean School, written around 100 CE.

Returning to Euclid's condition (5.6), when $p = 2, 3$, we get the perfect numbers 6 and 28, respectively, and when $p = 5, 7$ we get 496 and 8128, respectively.

In the *Introduction to Arithmetic*, Nicomachus further proposed five conjectures related to perfect numbers, the earliest and subsequently most famous conjectures about this topic. These are:

(1) the nth perfect number has n digits,
(2) every perfect number is even,
(3) the perfect numbers alternatingly end with the digits 6 and 8,
(4) the sufficient condition given in Euclid's *Elements* is also a necessary condition for an even number to be perfect, and
(5) there are infinitely many perfect numbers.

Among these, the first and third conjectures were later disproved, and the fourth was eventually proved by Euler. The second and fifth conjectures comprise what is known today as the perfect number problem: are there infinitely many perfect numbers, and are any of them odd? The name of Nicomachus deserves to be remembered if only for this alone, and his conjectures promoted the development not only of the theory of perfect numbers but even of number theory in general. Nicomachus was born in Gerasa, a Syrian province of the Roman Empire, now part of the Kingdom of Jordan.

The fifth perfect number appeared much later, after the Medieval years of the Dark Ages. It was discovered only in the 15th century, more precisely sometime between the years of 1456 and 1461, and its discoverer remains unknown. It is the eight digit number 33,550,336, corresponding to $p = 13$. This, however, was not enough to disprove the first conjecture above, because it could not be known if there were any perfect numbers missing from the list between then fourth and the fifth known perfect numbers.

In the year 1588, the Italian mathematician Petro Cataldi (1555–1626) discovered the sixth perfect number, 8,589,869,056, and the seventh perfect number, 137,438,691,328, corresponding respectively to $p = 17$ and $p = 19$. Although both the fifth and the sixth perfect numbers end in the digit 6, this was not enough to disprove the third of the Nicomachean conjectures, since it had still not yet been proved that condition (5.5) is not only sufficient but necessary for an even number to be perfect, and so it could not be ruled out that there might exist unknown perfect numbers between the known perfect numbers. This proof would not come for another 159 years.

In the same year that Cataldi discovered the sixth and seventh perfect numbers, the French Catholic priest and mathematician Marin Mersenne was born. Mersenne studied systematically integers of the form $M_n = 2^n - 1$, known later as Mersenne numbers; when a Mersenne number is prime, it is called a Mersenne prime. Clearly there are at least as many even perfect numbers as there are Mersenne primes, but it was not known at that time whether or not there were exactly as many.

When people had observed that $M_2 = 3, M_3 = 7, M_5 = 31, M_7 = 127$ are all prime, they naturally conjectured that every M_p with p prime is prime. However, the very next such Mersenne number disproves this conjecture:

$$M_{11} = 2^{11} - 1 = 2047 = 23 \times 89$$

and it is precisely the failure of this conjecture that provides suspense to the problem of perfect numbers.

In 1747, the Swiss mathematician Leonhard Euler, living in Berlin at the time, proved the fourth of Nichomachus's conjectures, specifically that any even perfect number must be of the form (5.6). Today, the proof is not difficult, but it took more than two thousand years to prove both parts of this necessary and sufficient condition, which is now known as the Euclid-Euler Theorem.

The Euclid–Euler Theorem. *An even number n is a perfect number if and only if*

$$n = 2^{p-1}(2^p - 1)$$

for some prime number p such that also $2^p - 1$ is prime.

According to this theorem, the determination of even perfect numbers is reduced to the determination of Mersenne primes. Today, the identification of Mersenne primes and correspondingly of perfect numbers has become an attractive problem in computer science. Whether or not there are infinitely many of them remains an immortal puzzle, the oldest, and perhaps the most difficult problem in the history of mathematics.

At this point, some 1700 years after Nicomachus had formulated them, the first and third of his conjectures were finally disproven, since the fifth perfect number has eight digits, and both the fifth and sixth perfect numbers end in the digit 6 (since there are no other prime numbers between 13 and 17). Since then, there has been no further speculation or conjecture about the digits of even perfect numbers.

On September 30 2017, the author posted the following on Sina Weibo:

> "A small discovery: the Parthenon in Athens is the model of classical beauty; its east and west sides have height and width nineteen meters and thirty-one meters respectively. The ratio of these two numbers is about $0.613\ldots$, not so far off from the golden ratio $0.618\ldots$. Recently, I happened to notice something about the perfect numbers, which were first introduced by the Pythagorean school. After 2500 years, we have discovered 49 perfect numbers; thirty of these terminate in the digit 6, and nineteen of them in the digit 8. I personally predict that also the fiftieth perfect number, when it is discovered, will be found to end with the digit 6. Moreover, if it turns out that there are infinitely many perfect numbers, perhaps the ratio between the number of them that terminate in 8 and the number of them that terminate in 6 will tend to the golden ratio."

A few days earlier, I had tallied up again the perfect numbers terminating in 6 and 8, respectively with the help of the table of even perfect numbers in an appendix to my book on the subject. I became familiar with them, and proceeded to check again the dimensions of the Parthenon in Athens, which led be to the above discovery and conjecture. It is not hard to prove that when the prime number p determining a Mersenne prime is either 2 or congruent to 1 modulo 4 , then the corresponding perfect number terminates in 6 , and when it is congruent to 3 modulo 4, then the corresponding perfect number terminates in 8.

At that time, I believed that the 50th Mersenne prime, and therefore also the 50th perfect number, would be discovered within 5 years. In fact, it was discovered only 3 months later by Jonathan Pace, an employee of a courier company based in Tennessee in the USA. This Mersenne prime is given by $p = 77232917$, and the corresponding perfect number terminates in the digit 6; therefore, we boldly propose the following conjecture.

Conjecture 5.6. *There are infinitely many even perfect numbers, and the ratio between the number of them terminating in 8 to the number of them terminating in 6 tends to the golden ratio.*

On December 7, 2018, an American enthusiast named Patrick Laroche discovered the 51st perfect number, which also terminates in the digit 6, or in other words corresponds to a Mersenne prime determined by a prime number congruent to 1 modulo 4. So far, the ratio between the number of prime numbers generating Mersenne primes of type $4m + 3$ to those of type $4m + 1$ remains 19:31, which is very close to the golden ratio.

Although prime numbers are equally distributed across the different sets of arithmetic progressions with equal common difference by Dirichlet's theorem on primes in arithmetic progressions, this no longer holds when we restrict the variable to prime values. In fact, from a statistical point of view, there are only half as many prime numbers of the form $4p + 1$ as there are primes of the form $4p + 3$ where p is required to be prime. In other words, where the prime numbers are more dense in an arithmetic progression with prime variable, those corresponding to Mersenne primes are more rare. Perhaps it is necessary to gain a better understanding of the distribution of prime numbers in arithmetic progressions with prime variables to gain insight into the secrets of the terminal digits of perfect numbers.

We frankly admit here that our conviction in the infinitude of even perfect numbers comes from this association with the golden ratio, which is an irrational number. Intriguingly, both the concepts of the perfect numbers and of the golden ratio were probably introduced by the Pythagoreans, but they did not know or even suspect that there might be some connection between the two.

Finally, we present another conjecture about perfect numbers and Mersenne primes.

Conjecture 5.7. *If p and $2p-1$ are both prime such that also $2^p - 1$ and $2^{2p-1} - 1$ are both prime, then p must be one of $2, 3, 7, 31$.*

A prime number p such that $2p + 1$ is also prime is known as a Sophie Germain prime. Therefore, Conjecture 5.7 states that there are only four primes which are similar to Sophie Germain prime but with addition replaced by subtraction that also generate Mersenne primes and, correspondingly, perfect numbers.

Since perfect numbers are exceedingly rare, number theorists throughout history have experimented with various generalizations of the concept. The most common of these requires that the sum of proper divisors of a number is a multiple of the number, in other words adding a coefficient k to the right-hand side of (5.5):

$$\sum_{d \,|\, n, d < n} d = kn.$$

A number n satisfying this equation is called a $(k + 1)$-*multiply perfect number* or a *perfect number of order* $k + 1$; when $k = 1$, a 2-multiply perfect number is the same thing as a perfect number. Mathematicians who have studied these numbers include Fibonacci (Fig. 5.3), Mersenne, Descartes, and Fermat, and later, Lehmer, Carmichael, and others (see Dickson, 2002). Some of them did not find any such numbers, while others were able to produce several, but the results were scattered and did not yield any clear criteria like the relationship between perfect numbers and Mersenne primes.

The first person to discover a perfect number of order $k + 1$ with $k > 1$ was the Welsh mathematician Robert Recorde (1512–1558), who discovered that 120 is a perfect number of order 3 in 1557; in the same book he introduced the symbol "=" for equality. Next was Fermat, who discovered that also 672 is 3-multiply perfect in 1637; in the same year, he proposed Fermat's last theorem. In 1644, Fermat also found an eleven digit 3-multiply perfect number. Prior to this,

Figure 5.3. Statue of Fibonacci.

Mersenne and Descartes had discovered 3-multiply perfect numbers with nine and ten digits, respectively. These three Frenchman also found $(k + 1)$-multiply perfect numbers for other values of k.

In the spring of 2012, the author introduced the concept of square sum perfect numbers, which are positive integers n satisfying the following identity:

$$\sum_{d \mid n, d < n} d^2 = 3n. \tag{5.7}$$

After investigating these numbers, we (Cai Tianxin, Chen Deyi and Zhang Yong) (see Cai *et al.*, 2015) obtained the following theorem.

Theorem 5.5. *All solutions to* (5.7) *are given by* $n = F_{2k-1}F_{2k+1}$ $(k \geq 1)$, *where* F_{2k-1} *and* F_{2k+1} *are twin Fibonacci primes.*

For this reason, we call the original perfect numbers M-perfect numbers due to their association with Mersenne primes, and the square sum perfect numbers F-perfect numbers, due to their relation to Fibonacci primes. Incidentally, the Japanese mathematician Kohji Matsumoto suggested calling them Yin and Yang perfect numbers at the 2013 Sino-Japanese Number Theory Conference in Fukuoka, because the English letters corresponding to Yin (female) and Yang (male) are F and M, respectively.

The largest Fibonacci prime known to date is F_{81839}, and the largest Fibonacci probable prime is $F_{1968721}$ (with 411439 digits). Among these, there are are only five pairs of twin Fibonacci primes, giving five F-perfect numbers $n = F_3 F_5, F_5 F_7, F_{11} F_{13}, F_{431} F_{433}, F_{569} F_{571}$ (note that the subscripts of a pair of twin Fibonacci primes must also be twin primes). Respectively, these are

10,
65,
20737,
7351080381692266976103362664212353326194801197040523391981
4585711917444519057612261963528801744523093107269516305744
10613670787 1525711296518385628509088429445930772087319647
4208257,
3523220957390444959595279062040480245884253791540018496569
5897596126849742246390276402878432136154463286879043721897
5172518365904797160002711185572855328278293823839001006460
4217978755993551604318057918269182928456761611403668577116
737601.

Among them, 10 is the only even F-perfect number, which is unsurprising, since there is only one even prime number, namely 2. The next possible F-perfect number must have at least 822878 digits, but we do not know if in fact a sixth F-perfect number, nor can we disprove that there might exist infinitely many of them.

We turn next to a more general situation; with a and b and given positive integers, we consider the equation

$$\sum_{d|n, d<n} d^a = bn. \tag{5.8}$$

We have the following theorem.

Theorem 5.6. *If* $a = 2, b \neq 3$ *or* $a \geq 3, b \geq 1$, *then* (5.8) *has at most finitely many solutions; in particular, if* $a = 2, b = 1$ *or* 2, *then* (5.8) *has no solutions.*

In other words, there are no more interesting perfect numbers of this type besides the M-perfect numbers and the F-perfect numbers. The proof of Theorems 5.5 and 5.6, required the introduction and proof of several lemmas.

In the spring of 2014, we also considered a more general form of square sum perfect numbers, namely positive integers n satisfying

$$\sum_{d \mid n, d < n} d^2 = L_{2s} n - F_{2s}^2 + 1,$$

where s is an arbitrary positive integer, and F_{2s} and L_{2s} are, respectively Fibonacci and Lucas numbers. We (Cai Tianxin, Wang Liuquan and Zhang Yong) (see Cai *et al.*, 2019) proved that apart from a finite number of computable exceptions, all numbers satisfying this identity are of the form

$$n = F_{2k+1} F_{2k+2s+1} \text{ or } F_{2k+1} F_{2k-2s-1}$$

where k is a positive integer, and F_{2k+1} and $F_{2k+2s+1}$ (or $F_{2k-2s-1}$) are Fibonacci primes. In particular, when $s = 1$, this is simply (5.5) and correspondingly Theorem 5.5.

Moreover, we proved that for any positive integer k, the equation

$$\sum_{d \mid n, d < n} d^2 = 2n + 4k^2 + 1$$

has infinitely many solutions if and only if de Polignac's conjecture is true, and in particular when $k = 1$, the equation

$$\sum_{d \mid n, d < n} d^2 = 2n + 5$$

has infinitely many solutions if and only if the twin prime conjecture is true.

We also discovered that for any integer k, the equation

$$\sum_{d \mid n, d < n} 2k \, d^2 - (2k - 1)d = (4k^2 + 1)n + 2$$

has infinitely many solutions if and only if there are infinitely many primes p such that $2kp + 1$ is also prime; in particular, when $k = 1$, this is the Sophie Germain prime conjecture. Similarly, there are infinitely many primes p such that $2kp - 1$ is also prime if and only if the equation

$$\sum_{d|n, d<n} 2k \, d^2 + (2k - 1)d = (4k^2 + 1)n + 4k$$

has infinitely many solutions. Here we observe that the physicist Albert Einstein (1879–1955) once remarked in his autobiographical notes that "the true laws cannot be linear, nor can they be derived from linearity...".

We also proved (see Cai *et al.*, 2015) the following theorem.

Theorem 5.7. *Let $n = pq$, where p and q are distinct prime numbers. If $n \mid \sigma_3(n) = \sum_{d|n, d<n} d^3$, then $n = 6$; let $n = 2^\alpha p(\alpha \geq 1)$ where p is an odd prime number. If $n \mid \sigma_3(n)$, then n is an even perfect number, and conversely (with the exception of 28).*

We proposed at the same time the following conjecture.

Conjecture 5.8. *If $n = p^\alpha q^\beta (\alpha \geq 1, \beta \geq 1)$, where p and q are distinct prime numbers, then $n \mid \sigma_3(n)$ if and only if n is an even perfect number other than 28.*

In 2018, Jiang Xingwang proved (see Jiang, 2018) that Conjecture 5.8 holds if $p = 2, q$ is an odd prime number, which gives the following theorem.

Theorem 5.8. *If $n = 2^\alpha q^\beta (\alpha \geq 1, \beta \geq 1)$, where q is an odd prime, then $n|^{\sigma_3}(n)$ if and only if n is an even perfect number other than 28.*

In 2019 we (Zhong Hao and Cai Tianxin) (see Zhong and Cai, 2017) proved that Conjecture 5.8 holds if $\alpha = 1$ and $q = 3$ or $q \equiv 2 \pmod 3$, which we state as the following theorem.

Theorem 5.9. *If $n = pq^\beta (\beta \geq 1)$, where p and q are distinct odd primes, and $q = 3$ or $q \equiv 2 \pmod 3$, then $n|^{\sigma_3(n)}$ if and only if n as an even perfect number other than 28.*

In 2020, Hung Viet Chu of the University of Illinois considered the case $k = 5$ and proved (see Chu, 2020) the following theorem.

Theorem 5.10. *If* $n = 2^\alpha q^\beta (\alpha \geq 1, \beta \geq 1)$, *then* $n | \sigma_5(n)$ *if and only if* n *is an even perfect number other than 496.*

In 2021, *Chu further proved* (*see Chu, 2021*) *the following.*

Theorem 5.11. *If* n *is an even perfect number, then* $n \mid \sigma_k(n)$ *for any odd number* $k > 1$ *if and only if* $n = 6$.

Of the four theorems mentioned above, Theorem 5.11 is the easiest to prove, as follows. We start by proving sufficiency. Suppose $k > 1$ is odd, $j \geq 0$. Then $\sigma_k(6) = 1^k + 2^k + 3^k + 6^k$.

Note that

$$3^k - 3 = 3\left(3^{k-1} - 1\right) \equiv 0(\mathrm{mod}\ 6); 2^k - 2 = 2(4^{(k-1)/2} - 1)$$
$$\equiv 0(\mathrm{mod}\ 6)$$

In other words, $3^k \equiv 3(\mathrm{mod}\ 6), 2^k \equiv 2(\mathrm{mod}\ 6)$, from which it follows that $6 \mid \sigma_k(6)$.

Next we consider the necessity. Let $n = 2^{p-1}(2^p - 1)$ where $p > 2$ is prime (so $n > 6$). We need to prove that $n \nmid \sigma_p(n)$. Since $\sigma_p(n)$ is a multiplicative function,

$$\sigma_p(n) = \sigma_p\left(2^{p-1}(2^p - 1)\right) = \left(1 + 2^p + \cdots + 2^{p(p-1)}\right)(1 + (2^p - 1)^p)$$

Suppose for the sake of contradiction that $n \mid \sigma_p(n)$, and note that $(2^{p-1}, 1 + (2^p - 1)^p) = 1$. We find that

$$2^p - 1 \mid 1 + 2^p + \cdots + 2^{p(p-1)} = \sum_{i=0}^{p-1} 2^{pi}.$$

Each term in the sum on the right is 1 modulo $2^p - 1$, so the sum is congruent to p modulo $2^p - 1$, contradicting the divisibility hypothesis. This completes the proof of Theorem 5.11.

In 2022, Wang Xiaoyu proved (see Wang, 2022) that if n is an even perfect number, then $n \nmid \sigma_k(n)$ for any even k.

Indeed, if $n = 6$, then

$$\sigma_k(6) = 1^k + 2^k + 3^k + 6^k \equiv \begin{cases} 1(\mathrm{mod}\ 3), & k = 0 \\ 2(\mathrm{mod}\ 3), & k > 1 \end{cases}$$

which shows that $6 \nmid \sigma_k(6)$. For $p > 2$, write

$$\sigma_k(n) = \sigma_k(2^{p-1}(2^p - 1)) = (1^k + 2^k + \cdots + 2^{k(p-1)})(1 + (2^p - 1)^k).$$

Suppose $n \mid \sigma_k(n)$, in other words $2^{p-1}(2^p - 1) \mid (1^k + 2^k + \cdots + 2^{k(p-1)})(1 + (2^p - 1)^k)$.

Noting that $(2^{p-1}, 1^k + 2^k + \cdots + 2^{k(p-1)}) = 1$, it follows that $2^{p-1} \mid 1 + (2^p - 1)^k$. But

$$1 + (2^p - 1)^k \, 1 + (2^p - 1)^k = 2 + \sum_{i=1}^{k} \binom{k}{i}(-1)^{k-i}2^{p-i} \equiv 2 \,(\mathrm{mod}\, 2^{p-1}),$$

contradiction!

5.4 *S*-Perfect Numbers

In fall of 2020, the American graduate student of Zhejiang University Tyler Ross introduced (see Ross, 2025) the *S*-perfect numbers and proved some results concerning them.

Definition 5.1. Let $S \subset \mathbb{Z}$ be any set of integers, and let $n \in \mathbb{Z}$ ($|n| > 1$). If there exist $\lambda_1, \ldots, \lambda_k \in S$ such that

$$1 + \sum_{j=1}^{k} \lambda_j d_j = n,$$

where $1 = d_0 < d_1 < \cdots < d_k < d_{k+1} = |n|$ are the positive divisors of n, then we call n an *S*-perfect number of the first kind; and if there exist $\lambda_0, \ldots, \lambda_k \in S$ such that

$$\lambda_0 + \sum_{j=1}^{k} \lambda_j d_j = n,$$

we call n an *S*-perfect number of the second kind.

Example 5.1. The $\{1\}$-perfect numbers of the first kind are the perfect numbers. The $\{0, 1\}$-perfect numbers of the second kind are called semiperfect numbers, numbers equal to the sum of some but not necessarily all of their proper divisors, such as $12 (= 2 + 4 + 6 = 1 + 2 + 3 + 6)$. In the following, we consider only *S*-perfect numbers of the first kind.

Example 5.2. When $S = \{-1, 1\}$, the first fifteen S-perfect numbers are $6, 12, 24, 28, 30, 40, 42, 48, 54, 56, 60, 66, 70, 78, 80, \ldots$.

The smallest odd $\{-1, 1\}$-perfect number is 945, which is also the smallest odd abundant number (that is, with divisor sum exceeding itself).

Definition 5.2. If n is an S-perfect number, we call the sum $n = 1 + \sum_{j=1}^{k} \lambda_j d_j$ (or $n = \lambda_0 + \sum_{j=1}^{k} \lambda_j d_j$) an S-presenation of n, or simply a presentation of n, when S is fixed.

The following result shows, that for most integers $|n| > 1$, it is easy to find $S \subset \mathbb{Z}$ such that n is S-perfect. For this reason, we focus on determining the S-perfect numbers and related properties when S is given.

Theorem 5.12. *If $n \in \mathbb{Z} (|n| > 1)$ has at least two prime factors, then there exists $S \subset \mathbb{Z}$ with cardinality $\#S \leq \tau(n) - 2$ such that n is S-perfect; if n is prime or a prime power, then n is not S-perfect for any $S \subset \mathbb{Z}$.*

Proof.   If n has at least two prime factors, say the divisors of n are

$$1 = d_0 < d_1 < \cdots < d_k < d_{k+1} = |n|$$

then $\gcd(d_1, \ldots, d_k) = 1$. It follows that the linear diophantine equation

$$\sum_{j=1}^{k} d_j x_j = n - 1$$

has solutions. The second claim is obvious. This completes the proof of Theorem 5.12. $\square$

Let $P(S)$ denote the set of S-perfect numbers for fixed S. If $(S_\alpha)_{\alpha \in A}$ is any family of sets of integers, it is easy to see that

$$\cup_{\alpha \in A} P(S_\alpha) \subset P(\cup_{\alpha \in A} S_\alpha), P(\cap_{\alpha \in A} S_\alpha) \subset \bigcap_{\alpha \in A} P(S_\alpha).$$

Next, we consider several special cases: $S = \{0, m\}(m \geq 1), S = \{-1, m\}(m \geq 1)$, and $S = \{-1, 1\}$. The simplest is $S = \{0, m\}$ (a variation on semiperfect numbers). The following lemma shows that there are infinitely many $\{0, m\}$-perfect numbers for all $m \geq 1$.

Lemma 5.3. *If* $n \in P(0, m)(m \geq 1)$, *then also* $(m + 1)n \in P(0, m)$.

Proof. If $n = \sum$ is a $\{0, m\}$-presentation of n, then $(m + 1)n = \sum + mn$ is a $\{0, m\}$-presentation of $(m + 1)n$. This proves the lemma.

According to Lemma 5.3, it suffices to find a single $n \in P(0, m)$, to show that there exist infinitely many such n. □

Theorem 5.13. *For every* $m \geq 1$, *there are infinitely many* $\{0, m\}$-*perfect numbers.*

Proof. Note that

$$(m + 1)(m^2 + m + 1) = 1 + m(m + 1) + m(m^2 + m + 1)$$

is $\{0, m\}$-perfect. By Lemma 5.3, this proves Theorem 5.13.

The $\{-1, m\}$-perfect numbers are more interesting. We consider only the $\{-1, m\}$-perfect numbers of the form $n = 2^k p$, where pp is an odd prime, $k \geq 1$. Write $\text{ord}_2(n)$ for the largest positive integer j such that 2^j divides n. □

Lemma 5.4. *If* $0 \leq s \leq t, m \geq 1$, *then* $n = \sum_{j=s}^{t} \lambda_j \cdot 2^j$ *for some* $\lambda_s, \ldots, \lambda_t \in \{-1, m\}$ *if and only if*

$$-\sum_{j=s}^{t} 2^j = -2^s(2^{t-s+1} - 1) \leq n \leq 2^s m(2^{t-s+1} - 1) = \sum_{j=s}^{t} 2^j m$$

and

$$n \equiv -2^s(2^{t-s+1} - 1)(\bmod 2^s(m + 1)).$$

Proof. For any $\lambda_s, \ldots, \lambda_t \in \{-1, m\}$, it is easy to see that

$$\sum_{j=s}^{t} \lambda_j \cdot 2^j \equiv -\sum_{j=s}^{t} 2^j (\bmod 2^s(m + 1))$$

and these numbers are distinct for every distinct choice of $\lambda_s, \ldots, \lambda_t \in \{-1, m\}$. A simple counting argument completes the lemma. □

Lemma 5.5. *Let* $m \geq 1, \beta = \text{ord}_2(m + 1)$. *If* $n = \sum_{j=s}^{t} \lambda_j \cdot 2^j$ *for some* $0 \leq s \leq t$, $\lambda_s, \ldots, \lambda_t \in \{-1, m\}$, *with* $t \geq s + \beta - 1$, *then there exist also some* $\Lambda_s, \ldots, \Lambda_{t+\alpha} \in \{-1, m\}$ *such that* $n = \sum_{j=s}^{t+\alpha} \Lambda_j \cdot 2^j$ *whenever* $2^\alpha \equiv 1(\bmod (m + 1)/2^\beta)$.

Proof. If $2^\alpha \equiv 1 (\mathrm{mod}\, (m+1)/2^\beta)$, then $2^{t+\alpha+1} \equiv 2^{t+1}(\mathrm{mod}\, 2^{t+1-\beta}(m+1))$. Also, if $t+1-\beta \ge s$, then $2^{t+\alpha+1} \equiv 2^{t+1}(\mathrm{mod}\, 2^s(m+1))$. Therefore,

$$-2^s \left(2^{t-s+1} - 1\right) \equiv -2^s(2^{t+\alpha-s+1} - 1)(\mathrm{mod}\, 2^s(m+1))$$

as required by Lemma 5.4 This completes the proof of Lemma 5.5.

$\square$

Theorem 5.14. *Let* $m \ge 1, \beta = \mathrm{ord}_2(m+1)$. *If for any odd prime* p *and any* $k \ge 1$, *both* $2^k p$ *and* $2^{k+\alpha} p$ *are* $\{-1, m\}$-*perfect, then* $2^\alpha \equiv 1 \left(\mathrm{mod}\, (m+1)/2^\beta\right)$. *Conversely, if* $2^k p \in P(-1, m)$ *for some prime* p *and some* $k \ge \beta$, *then* $2^{k+\alpha} p \in P(-1, m)$ *whenever* $2^\alpha \equiv 1 \left(\mathrm{mod}\, (m+1)/2^\beta\right)$.

Proof. Suppose first that both $2^k p$ and $2^{k+\alpha} p \in P(-1, m)$, with presentations

$$2^k p = 1 + \sum_{j=1}^{k} \lambda_j^{(1)} \cdot 2^j + \sum_{j=0}^{k-1} \lambda_j^{(2)} \cdot 2^j p \qquad (5.9)$$

$$2^{k+\alpha} p = 1 + \sum_{j=1}^{k+\alpha} \Lambda_j^{(1)} \cdot 2^j + \sum_{j=0}^{k+\alpha-1} \Lambda_j^{(2)} \cdot 2^j p, \qquad (5.10)$$

respectively. Note that every $\lambda_j^{(i)}, \Lambda_j^{(i)} \equiv -1(\mathrm{mod}\, m+1)$. Then (5.9) gives

$$2^k p \equiv 1 - \sum_{j=1}^{k} 2^j - \sum_{j=0}^{k-1} 2^j p(\mathrm{mod}\, m+1)$$

or

$$\left(2^{k+1} - 1\right)(p+1) \equiv 2(\mathrm{mod}\, m+1)$$

from which it follows easily that $p+1$ must be a unit modulo $(m+1)/2^\beta$.

Subtracting (5.9) from (5.10) and reducing again modulo $m+1$, we get

$$2^k p(2^\alpha - 1) \equiv -\sum_{j=k+1}^{k+\alpha} 2^j - \sum_{j=k}^{k+\alpha-1} 2^j p(\mathrm{mod}\, m+1)$$

or

$$2^{k+1}(p+1)(2^\alpha - 1) \equiv 0(\mathrm{mod}\, m + 1)$$

which implies that also $2^{k+1}(p+1)(2^\alpha - 1) \equiv 0 \left(\mathrm{mod}\, (m+1)/2^\beta\right)$. Since both $2^{k+1}2^{k+1}$ and $p+1$ are units modulo $(m+1)/2^\beta$, we conclude that $2^\alpha - 1 \equiv 0(\mathrm{mod}\, (m+1)/2^\beta)$.

Conversely, suppose that $k \geq \beta k \geq \beta$ and $2^k p \in P(-1, m)$ for some odd prime p, with presentation given by (5.9). Suppose also that $2^\alpha \equiv 1(\mathrm{mod}\, (m+1)/2^\beta)$. Then

$$2^{k+\alpha}p = 1 + \sum_{j=1}^{k} \lambda_j^{(1)} \cdot 2^j + \sum_{j=0}^{k-1} \lambda_j^{(2)} \cdot 2^j p + \sum_{j=k}^{k+\alpha-1} 2^j p. \qquad (5.11)$$

Since $k \geq \beta$, Lemma 5.5 applies, and we can find $\Lambda_1^{(1)}, \ldots, \Lambda_{k+\alpha}^{(1)}$ such that

$$\sum_{j=1}^{k+\alpha} \Lambda_j^{(1)} \cdot 2^j = \sum_{j=1}^{k} \lambda_j^{(1)} \cdot 2^j.$$

As for the remaining sum in (5.11), set $A = \sum_{j=0}^{k-1} \lambda_j^{(2)} \cdot 2^j + \sum_{j=k}^{k+\alpha-1} 2^j$. Reducing modulo $m + 1$,

$$A \equiv 2^{k+\alpha} - 2^{k+1} + 1 \equiv -(2^{k+\alpha} - 1)(\mathrm{mod}\, m + 1),$$

where we have made use of the hypotheses $2^\alpha \equiv 1(\mathrm{mod}\, (m+1)/2^\beta)$ and $k \geq \beta$ to substitute $2^{k+\alpha} \equiv 2^k (\mathrm{mod}\, m + 1)$. Then A satisfies the conditions of Lemma 5.4 (with $s = 0, t = k + \alpha - 1$), so we can find $\Lambda_0^{(2)}, \ldots, \Lambda_{k+\alpha-1}^{(2)}$ such that $A = \sum_{j=0}^{k+\alpha-1} \Lambda_j^{(2)}$. We obtain a presentation

$$2^{k+\alpha}p = 1 + \sum_{j=1}^{k+\alpha} \Lambda_j^{(1)} \cdot 2^j + \sum_{j=0}^{k+\alpha-1} \Lambda_j^{(2)} \cdot 2^j p.$$

This completes the proof of Theorem 5.14.   □

It follows that it is sufficient to find a single $2^k p \in P(-1, m)$ (p an odd prime, $k \geq \beta$) to generate infinitely many. The following theorem gives a construction.

Theorem 5.15. *Let $m \geq 1$, $\beta = \mathrm{ord}_2(m+1)$. Take $\alpha > \beta$ such that $2^\alpha \equiv 1 (\mathrm{mod}\,(m+1)/2^\beta)$. If $p \equiv 2(2^{\alpha+1}-1)-1 (\mathrm{mod}\,2(m+1))$ is prime, then there exists $k \geq \alpha$ such that $2^k p \in P(-1,m)$.*

Proof. Set $N = 2^{\alpha+1}-1$, and note that $\alpha > \beta$ implies that $N^2 \equiv 1(\mathrm{mod}\,2(m+1))$. If $p \equiv 2\left(2^{\alpha+1}-1\right)-1(\mathrm{mod}\,2(m+1))$, then

$$Np \equiv 2N^2 - N = 2 - N = 3 - 2^{\alpha+1}(\mathrm{mod}\,2(m+1)).$$

In other words, $Np-1 \equiv -2\left(2^\alpha - 1\right)(\mathrm{mod}\,2(m+1))$. Take $k \geq \alpha$ such that $2^k \equiv 2^\alpha(\mathrm{mod}\,2(m+1))$ and $\quad Np-1 \leq 2m\left(2^k - 1\right)$. By Lemma 5.4, there exist $\lambda_1^{(1)},\ldots,\lambda_k^{(1)} \in \{-1,m\}$ such that

$$Np = 1 + \sum_{j=1}^{k} \lambda_j^{(1)} \cdot 2^j.$$

We also have $2^k - N \equiv -\left(2^k - 1\right)(\mathrm{mod}\,2(m+1))2^k - N \equiv -(2^k - 1)(\mathrm{mod}\,2(m+1))$, so there exist $\lambda_1^{(2)},\ldots,\lambda_{k-1}^{(2)} \in \{-1,m\}$ such that

$$2^k - N = \sum_{j=0}^{k-1} \lambda_j^{(2)} \cdot 2^j.$$

We get a presentation

$$1 + \sum_{j=1}^{k} \lambda_j^{(1)} \cdot 2^j + \sum_{j=0}^{k-1} \lambda_j^{(2)} \cdot 2^j p = Np + (2^k - N)p = 2^k p.$$

This completes the proof of Theorem 5.15.   □

Corollary 5.1. *There exist infinitely many $\{-1,m\}$-perfect numbers for every positive integer m.*

Proof. With α, β as in Theorem 5.15, $2(2^{\alpha+1}-1)-1 \equiv 1(\mathrm{mod}\,(m+1)/2^\beta)$; since $2(2^{\alpha+1}-1)-1$ is odd, we find that $gcd(2(2^{\alpha+1}-1)-1, 2(m+1)) = 1$. It follows by Dirichlet's theorem on primes in arithmetic progressions that there do in fact exist primes $p \equiv 2(2^{\alpha+1}-1)-1(\mathrm{mod}\,2(m+1))$. This proves the corollary.   □

Example 5.3. We use Theorems 5.10 and 5.11 to construct some presentations. Consider $m = 23$, which gives $m + 1 = 24.2(m + 1) = 48$, $\beta = 3$, $(m + 1)/2^\beta = 3$; then $\alpha = 4$ is the smallest $\alpha \geq \beta$ such that $2^\alpha \equiv 1 (\mathrm{mod}\,(m + 1)/2^\beta)$. Setting $N = 2^{\alpha+1} - 1 = 31, 2N - 1 \equiv 13 (\mathrm{mod}\,2(m + 1))$, we can take $p = 13$. Since $690 = 2m\,(2^4 - 1) > Np - 1 = 402$, then according to Theorem 5.15, $208 = 2^4 \cdot 13 \in P(-1, 23)$. And indeed, considering $Np = 403$ and $2^4 - N = -15$, then following Lemma 5.4 we find that

$$403 = 1 + 23 \cdot 2 - 2^2 - 2^3 + 23 \cdot 2^4, \quad -15 = -1 - 2 - 2^2 - 2^3.$$

This gives the presentation

$$208 = 1 + 23(2) - 2^2 - 2^3 + 23\,(2^4) - 13 - 2 \cdot 13 - 2^2 \cdot 13 - 2^3 \cdot 13.$$

Next, since $2^2 \equiv 1 (\mathrm{mod}\,(m + 1)/2^\beta)$, Theorem 5.14 implies that $403 = 2^4 \cdot 13, 832 = 2^6 \cdot 13, 3328 = 2^8 \cdot 13, 13312 = 2^{10} \cdot 13 \cdots \in P(-1, 23)$ (in fact, this follows already from Theorem 5.15; Theorem 5.14 is really only necessary if the first number in the sequence is not obtained from Theorem 5.15). We verify only that $2^6 \cdot 13 \in P(-1, 23)$. Examining the presentation above of $208 = 2^4 \cdot 13$, we have $1 + 23(2) - 2^2 - 2^3 + 23\,(2^4) = 403$ and $-1 - 2 - 2^2 - 2^3 + 2^4 + 2^5 = 33$. Then following the proof of Theorem 5.15, we find that

$$403 = 1 + 23 \cdot 2 + 23 \cdot 2^2 - 2^3 + 23 \cdot 2^4 - 2^5 - 2^6,$$
$$33 = -1 - 2 + 23 \cdot 2^2 - 2^3 - 2^4 - 2^5.$$

From this we get the presentation

$$832 = 1 + 23(2) + 23(2^2) - 2^3 + 23(2^4) - 2^5 - 2^6 - 13 - 2 \cdot 13$$
$$+ 23(2^2 \cdot 13) - 2^3 - 2^4 \cdot 13 - 2^5 \cdot 13.$$

Finally, we consider in more detail the case $m = 1$ and obtain several results special to the $\{-1, 1\}$-perfect numbers, which possess a certain aesthetic appeal due to the formal similarity of the sum involved to the divisor sum in the definition of perfect numbers.

We first refine slightly the relevant special case of Lemma 5.4.

Lemma 5.6. *For any integer n, there exist $k \geq 1$ and $\lambda_1, \ldots, \lambda_k \in \{-1, 1\}$, such that $n = 1 + \sum_{j=1}^{k} \lambda_j \cdot 2^j$ if and only if $n \equiv 3 (\mathrm{mod}\,4)$.*

Proof. Take $k \geq 1$ such that $-2(2^k - 1) \leq n - 1 \leq 2(2^k - 1)$. Then the conditions of Lemma 5.4 are satisfied (with $m = 1, s = 1, t = k$). This completes the proof of Lemma 5.6. □

Lemma 5.7. *Let n be any integer, p a prime number not dividing n. Then (1) if $n \in P(-1, 1)$ $n \in P(-1, 1)$, then also $np^k \in P(-1, 1)$ for all $k \geq 1$, and (2) if $np \in P(-1, 1)$, then also $np^{2k-1} \in P(-1, 1)$ for all $k \geq 1$.*

Proof. If (1) $n = \Sigma_1$ and $np^k = \Sigma_2(k \geq 0)$ are presentations of n and np^k, respectively, then $np^{k+1} = \Sigma_2 - np^k + p^{k+1}\Sigma_1$ is a presentation of np^{k+1}. Similarly, if (2) $np = \Sigma_1$ $np = \Sigma_1$ and $np^k = \Sigma_2$ $(k \geq 1)$ are presentations of np and np^k, respectively, then $np^{k+2} = \Sigma_2 - np^k + p^{k+1}\Sigma_1$ is a presentation of np^{k+2}. This completes the proof of Lemma 5.7. □

Lemma 5.8. *If any $n \in P(-1, 1)$, then also $2n \in P(-1, 1)$.*

Proof. If n is odd, this follows immediately from Lemma 5.7. Suppose n is even, and $n = 1 + \sum_{j=1}^{k} \lambda_j d_j$ is a presentation of n. Then $2n = 1 + \sum_{j=1}^{k} \lambda_j d_j + n$. The proper divisors of $2n$ missing from this sum are of the form $2d_j$, where d_j divides n and $1 < d_j < n$ (because n is even). Replace every such $\lambda_j d_j$, in the sum with $-\lambda_j d_j + \lambda_j (2d_j)$ to obtain a presentation of $2n$. This completes the proof of Lemma 5.8. □

Theorem 5.16. *If $d \geq 1$ is odd and not a square, then $2^k d \in P(-1, 1)$ for all but finitely many $k \geq 1$. Conversely, if $2^k d \in P(-1, 1)$ for some $k \geq 0, d \geq 1$, then d is not a square.*

Proof. In light of Lemmas 5.7 and 5.8, it suffices to show that $2^k p \in P(-1, 1)$ for some $k \geq 1$ for every odd prime p. Choose (Lemma 5.6) $k \geq 1$ and $\lambda_1, \ldots, \lambda_k$ such that

$$1 + \sum_{j=1}^{k} \lambda_j 2^j = \begin{cases} p, & p \equiv 3 \pmod 4, \\ 3p, & p \equiv 1 \pmod 4. \end{cases}$$

Then

$$2^k p = 1 + \sum_{j=1}^{k} \lambda_j 2^j + (-1)^{(p+1)/2} p + \sum_{j=1}^{k-1} 2^j p$$

is a presentation.

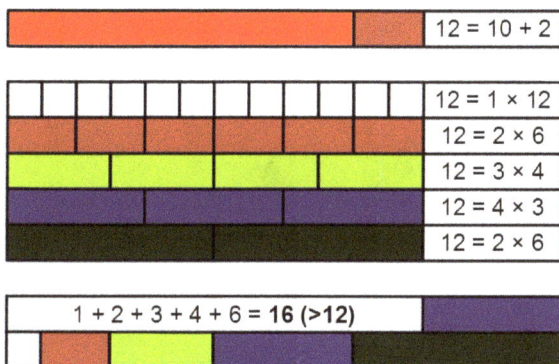

Figure 5.4. Depiction of the abundant number 12 with colored blocks.

Conversely, if $n \in P(-1, 1)$ with presentation $n = 1 + \sum_{j=1}^{k} \lambda_j d_j$, then

$$\sigma(n) = \sum_{j=1}^{k}(1 - \lambda_j)d_j + 2n$$

is even (since every $1 - \lambda_j = 0$ or 2). But it is well known that $\sigma(n)$ is even if and only if n is not a square or twice a square. This completes the proof of Theorem 5.16.

We add here a few further conjectures and questions concerning $\{-1, 1\}$-perfect numbers. Obviously, every $\{-1, 1\}$-perfect number is an abundant number. On the other hand, not every abundant number is $\{-1, 1\}$-perfect. The first few abundant numbers that are not also $\{-1, 1\}$-perfect are $18, 20, 36, 72, \ldots$. We conjecture that almost every abundant number is $\{-1, 1\}$-perfect (Fig. 5.4). □

Conjecture 5.9. *The positive $\{-1, 1\}$-perfect numbers have a natural density, equal to the natural density A of the abundant numbers. In 1998, Mark Deléglise proved (see Deléglise, 1998) that $0.2474 < A < 0.2480$.*

It is well known that the smallest odd abundant number is 945, which is also $\{-1, 1\}$-perfect, as, in fact, is every other odd abundant number smaller than 10^4. On the other hand, theoretical considers show that not every odd abundant number is $\{-1, 1\}$-perfect: for example, if n is an odd abundant number, then so is n^2, since the abundant numbers are closed under multiplication; but Theorem 5.12 shows that n^2 cannot be $\{-1, 1\}$-perfect. We make the following conjecture.

Conjecture 5.10. *Every non-square odd abundant number is $\{-1.1\}$-perfect.*

Chapter 6

The *abcd* Equation and New Congruent Numbers

Waiting slowly becomes a poem in the gap of time.

— *The Loneliness of Primes*

6.1 The *abcd* Equation

In 2013, about a year after introducing the F-perfect numbers, the author defined (see Cai, 2021) the *abcd* equation, given in the following, which unexpectedly turned out to also be closely related to the Fibonacci sequence. The methods used in the study of this topic are rich, and the difficulty of solving it completely is immense. The problems of judging whether or not there exists any solution and determining the number and structure of the solutions when they exist, for example whether or not there are infinitely many, are all worthwhile issues for exploration.

Definition 6.1. Let n be a positive integer, a, b, c, d positive rational numbers, then the *abcd* equation is

$$n = (a + b)(c + d), \tag{6.1}$$

where

$$abcd = 1.$$

By the arithmetic-geometric inequality, $(a + b)(c + d) \geq 2\sqrt{ab} \times 2\sqrt{cd} = 4$, so (6.1) has no solutions when $n = 1, 2$, or 3. On the other hand, $4 = (1 + 1)(1 + 1)$, $5 = (1 + 1)(2 + \frac{1}{2})$.

It is easy to see that if (6.1) admits positive rational solutions, then also the equation

$$n = x + \frac{1}{x} + y + \frac{1}{y} \tag{6.2}$$

admits positive rational solutions; the converse is also true, since

$$x + \frac{1}{x} + y + \frac{1}{y} = (x + y)\left(1 + \frac{1}{xy}\right).$$

It is also not difficult to see that every solution to (6.2) corresponds to an infinite family of solutions $(ka, kb, \frac{c}{k}, \frac{d}{k})$ to (6.1). In particular, when $n = 4$ or 5, (6.2) has unique solutions $(x, y) = (1, 1)$ and $(x, y) = (2, 2)$, respectively. The first claim is obvious, but the second requires the use of the theory of elliptic curves to prove.

First, we have the following criterion.

Criterion 6.1. If $8|n$ or $2\|n$, then the *abcd* equation has no solutions; if n is odd or $4\|n$ and moreover n has prime factors congruent to 3 modulo 4, then also the *abcd* equation has no solutions.

Proof. If (6.2) has a solution, then there exist positive integers a_1, b_1, c_1, d_1 such that

$$n = \frac{a_1}{b_1} + \frac{b_1}{a_1} + \frac{c_1}{d_1} + \frac{d_1}{c_1}, (a_1, b_1) = (c_1, d_1) = 1.$$

Multiplying both sides by $a_1 b_1$, we get

$$a_1 b_1 n - (a_1^2 + b_1^2) = \frac{a_1 b_1}{c_1 d_1}(c_1^2 + d_1^2).$$

Since $(c_1 d_1, c_1^2 + d_1^2) = 1$, we must have that $c_1 d_1 | a_1 b_1$, since otherwise the right-hand side is not an integer. Similarly, $a_1 b_1 | c_1 d_1$; so $a_1 b_1 = c_1 d_1$. $\square$

We also have

$$a_1 b_1 n = a_1^2 + b_1^2 + c_1^2 + d_1^2 = (a_1 \pm b_1)^2 + (c_1 \mp d_1)^2. \tag{6.3}$$

Considering parity, symmetry, and the fact that $a_1 b_1 = c_1 d_1$, we can distinguish two possible cases for (a_1, b_1, c_1, d_1), namely (odd, odd,

odd, odd) and (odd, even, odd, even). Since the square of any even number modulo 4 is zero and the square of any odd number modulo 8 is 1, we conclude that (6.3) has no solutions, and therefore also the *abcd* equation has no solutions, when $8|n$ or $2||n$.

When n is an odd number or $4||n$, the theory of quadratic residues shows that the left side of (6.3) and n in particular cannot have any prime factors congruent to 3 modulo 4. Otherwise, considering $p \equiv 3(\bmod 4)$, if $p|c_1 + d_1$, $p|c_1 - d_1$, then $p|(c_1, d_1)$, contradiction. But if $p \nmid c_1 \pm d_1$, then $(\frac{-(c_1 \pm d_1)^2}{p}) = -1$, where $()$ is the Legendre symbol, also a contradiction.

Next, we consider the equation

$$n = \left(a + \frac{1}{a} \right) \left(b + \frac{1}{b} \right) \tag{6.4}$$

with both a and b positive integers. It is obvious that the *abcd* equation has equations if (6.4) does, and (6.4) has a solution if and only if $(a, b) = 1$ and

$$a|b^2 + 1, \quad b|a^2 + 1.$$

It is not hard to see that the above condition is equivalent to

$$a^2 + b^2 + 1 \equiv 0(\bmod ab),$$

or in other words equivalent to the existence of a positive integer q such that

$$a^2 + b^2 + 1 = qab.$$

It is possible to prove that this equation (in variables a, b) has a solution if and only if $q = 3$, in which case all solutions to (6.4) are of the form $a = F_{2k-1}$, $b = F_{2k+1}$.

Then from the identity

$$F_{n-1}F_{n+1} - F_n^2 = (-1)^n (n \geq 1),$$

between Fibonacci numbers, known as Cassini's identity, we get the following theorem.

Theorem 6.1. *The abcd equation has a solution, with* $(a, b) = (F_{2k-1}, F_{2k+1})$, *whenever* $n = F_{2k-3}F_{2k+3}(k \geq 0)$.

It follows from Theorem 6.1 that there are infinitely many n $(4, 5, 13, 68, 445, 3029, 20740, \ldots)$ such that the *abcd* equation has a solution.

Moreover, using properties of the Pisano period, we can prove the following theorem.

Theorem 6.2. *If n is an odd number such that (6.4) has a solution, then necessarily $n = 5 \pmod 8$; if n is an even number such that (6.4) has a solution, then necessarily $n = 4m$, $m = 1 \pmod{16}$.*

Proof. If n is such that (6.4) has a solution, then Theorem 6.1 shows that $n = F_{2k-3} F_{2k+3}$. The Pisano cycle of the Fibonacci sequence modulo 8 is $\{1, 1, 2, 3, 5, 0, 5, 5, 2, 7, 1, 0\}$, with a length of 12. Therefore, the values of n modulo 8 have a cycle of length 6. When $k = 3$ or 6, n is even; when $k = 1, 2, 4$, or 5, $n = 5 \pmod 8$. It is easy to see then that if n is an even number such that (6.4) has a solution, then $n = F_{6k-3} F_{6k+3}$. Since the Fibonacci sequence satisfies $F_{6s+3} \equiv 2 \pmod{32}$, that is, it is of the form $32k + 2$, after multiplying these two numbers together we must have $n = 4 \pmod{64}$. This completes the proof of Theorem 6.2. □

We turn next to the equation

$$n = \left(\frac{a}{b} + \frac{b}{a} \right) \left(\frac{c}{d} + \frac{d}{c} \right), \tag{6.5}$$

where a, b, c, d are positive integers with $(a, b) = (c, d) = 1$.

Obviously, (6.4) is a special case of (6.5). If (6.5) has a solution, then so do both (6.1) and (6.2); the converse can be shown via transformations and congruences. We refer to any of (6.1), (6.2), and (6.5) collectively as the *abcd* equation.

We have obtained only a few results so far for (6.5). For example, when $b = 1$, we find the following solution: if $2c | a^2 + 1$, $a | c^2 + 4$, there are solutions

$$1237 = (17/1 + 1/17)(145/2 + 2/145),$$
$$6925 = (337/1 + 1/337)(41/2 + 2/41),$$

where 1237 is a prime number; when a, c, d are odd, and $cd|a^2 + 1$, $a|c^2 + d^2$, there are solutions

$$580 = (157/1 + 1/157)(5/17 + 17/5),$$
$$1156 = (73/1 + 1/73)(13/205 + 205/13),$$
$$5252 = (697/1 + 1/697)(5/37 + 37/5).$$

More interestingly, we can obtain infinitely many solutions to (6.4) or (6.5), that is, the $abcd$ equation; for example, with $(c, d) = (1, 1)$, $(41, 137)$, $(386, 35521)$, each pair generates a sequence in which any two adjacent numbers produces a solution to (6.5). Setting $a = c^2 + d^2$, $b = 1$, then (a, b, c, d) corresponds to

$$n = \{(c^2 + d^2)^2 + 1\}/cd.$$

The first three such sequences are

$$\ldots\ldots 41761, 17, 2, 1, 1, 2, 17, 41761, \ldots,$$
$$\ldots\ldots 20626, 41, 137, 8592082, \ldots,$$
$$\ldots\ldots 624977, 386, 35531, \ldots,$$

where each adjacent triple (a, b, c) satisfies $ac = b^4 + 1$.

On the basis of Criterion 6.1, we also prove the following criterion.

Criterion 6.2. Under the hypotheses of the second part of Criterion 6.1, if $n \pm 4$ has a prime factor $p \equiv 3 \pmod 4$, then necessarily $p^{2k} \| n \pm 4$ for some positive integer k.

Proof. From the proof of Criterion 6.1, we know that if (6.2) has a solution, then there exist positive integers a, b, c, d with $(a, b) = (c, d) = 1$, $ab = cd$ such that

$$nab = (a \pm b)^2 + (c \mp d)^2.$$

By the theory of quadratic residues, it is easy to see that a, b cannot have any prime factors congruent to 3 modulo 4. Moving terms around, we have

$$(n \pm 4)ab = (a \pm b)^2 + (c \pm d)^2. \tag{6.6}$$

Suppose $p \equiv 3 \pmod 4$ is prime and $p|n + 4$; if $p \nmid c + d$, then again by the theory of quadratic residues it is clear that (6.6) cannot hold;

if $p|c+d$, then also $p|a+b$, so $p^2|n+4$. Also if $p^k|n+4$, $k > 2$, divide both sides of (6.6) by p^2, and repeating the argument we find that $p^{2k}|n+4$. We prove similarly that if $p \equiv 3(\text{mod } 4)$ and $p|n-4$, then $p^{2k}||n-4$. This completes the proof. $\qquad\square$

Corollary 6.1. *For any nonnegative integer k, if $n = F_{2k-3}F_{2k+3}$, then either n is odd or else $4||n$, and n has no prime factors congruent to 3 modulo 3; moreover, if $n \pm 4$ has a prime factor $p \equiv 3(\text{mod } 4)$, then $p^{2k}||n \pm 4$ for some positive integer k.*

Corollary 6.2. *If $n = 4m$ for some integer m such that the abcd equation has solutions, then necessarily $m = 1(\text{mod } 8)$.*

Proof of Corollary 6.2. According to Criterion 6.1, in this case $m \equiv 1(\text{mod } 4)$. If m is of the form $m = 8k + 5$ for some integer k, then $n + 4 = 8(4k + 3)$, which implies that $n + 4$ must have a prime factor of the type $4j + 3$. Then Criterion 6.2 implies that the abcd equation has no solutions. Therefore, we must have $m \equiv 1(\text{mod } 8)$.

Considering Criterions 6.1 and 6.2 and Theorem 6.2, we can conclude that apart from $n = 4, 5, 13, 68, 445$, and 580, for which the abcd equation has solutions, the remaining positive integers not exceeding 1000 such that the abcd equation could possibly admit solutions are $n = 41, 85, 113, 149, 229, 265, 292, 365, 373, 401, 481, 545, 761, 769, 797, 877, 905, 932$.

Following upon our calculations and analyses, we make the following conjectures.

Conjecture 6.1. *If $n = 1(\text{mod } 8)$ is a positive integer, then the abcd equation has no solutions.*

Conjecture 6.2. *If $n = 4m$ such that the abcd equation has a solution, then necessarily $m = 1(\text{mod } 16)$.*

Assuming the two conjectures above to be true, this narrows down the remaining possible positive integers not exceeding 1000 such that the abcd equation might have a solution to $n = 85, 149, 229, 365, 373, 797, 877$.

Figure 6.1. Institut Henri Poincaré, photograph by the author in Paris.

Among these, we find a solution with two positive and two negative terms for $n = 149$:

$$149 = \left(\frac{14640}{91} + \frac{91}{14640} - \frac{3965}{336} \right) \left(\frac{91}{14640} - \frac{3965}{336} - \frac{336}{3965} \right)$$

$$= \left(\frac{14640}{91} - \frac{3965}{336} \right) \left(1 - \frac{91 \times 336}{14640 \times 3965} \right).$$

In addition to these two conjectures, we also raise the following questions.

Question 6.1. Are there infinitely many positive integers n such that the *abcd* equation admits a solution but (6.4) does not?

Question 6.2. Are there infinitely many prime numbers n such that the *abcd* equation has solutions?

Question 6.3. Does there exist any positive integer n satisfying $n = (a+b)(c+d)$, $abcd = 1$, $ab \neq k^2$?

Remarks. We can obtain a generalization of the *abcd* equation by replacing the condition that $abcd = 1$ with instead the condition that $abcd = k^2$, leaving the other conditions in the definition unchanged. In this case, the necessary and sufficient conditions for the generalized *abcd* equation to have a solution remain valid in part, another new problem worth exploring.

6.2 Rational Point Constructions

In this section, we consider the number of solutions to (6.2) when such solutions exist. From the arithmetic-geometric inequality, it is easy to see that (6.2) has a unique solution when $n = 4$; if $n > 4$, we will need to transform (6.2) into an elliptic curve.

Theorem 6.3. *For $n > 4$, (6.2) has a solution if and only if the elliptic curve*

$$E_n : Y^2 = X^3 + (n^2 - 8)X^2 + 16X$$

has a rational point with $X < 0$.

Proof. Suppose $x \geq 1$, $y \geq 1$, $x + y > 2$, and consider the transformation

$$\begin{cases} x = \frac{s+nt}{2(t+t^2)} \\ y = \frac{s+nt}{2(1+t)}, \end{cases} \quad s, t > 0. \tag{6.7}$$

Its inverse transformation is given by

$$\begin{cases} t = \frac{y}{x}, \\ s = \frac{2y^2 + (2x-n)y}{x}. \end{cases}$$

showing that (6.7) defines a bijection. Substituting into (6.2), we get

$$n = \frac{(s + nt)^2 + 4t(1 + t^2)^2}{2t(s + nt)}.$$

Simplifying and putting

$$\begin{cases} X = -4t \\ Y = 4s, \end{cases}$$

we obtain E_n. This completes the proof of Theorem 6.3. □

We see also that here

$$\begin{cases} x = \frac{2Y - 2nX}{X^2 - 4X}, \\ y = \frac{Y - nX}{2(4 - X)}. \end{cases} \tag{6.8}$$

Example 6.1. The equation

$$5 = x + \frac{1}{x} + y + \frac{1}{y}$$

has a unique solution in positive rational numbers, given by $x = y = 2$.

Verification. According to Theorem 6.3, we need only to find a solution to the elliptic curve

$$E_4 : Y^2 = X^3 + 17X^2 + 16X, \, X < 0.$$

Using the Magma package, we find that E_5 has rank 0. By Mordell's theorem, the set $E_k(\mathbb{Q})$ of all rational points of the elliptic curve E_5 over the field of rational numbers is a finitely generated abelian group, given by

$$E_5(\mathbf{Q}) \cong T \bigoplus Z^{\mathrm{rank}(E_5)},$$

where T is the torsion part of $E_5(\mathbb{Q})$, which can be calculated directly as

$$T = \{(-16.0), (-4, -12), (-4, 12), (-1, 0), (0, 1), (4, -20), (4, 20), \infty\},$$

the last of these being the point at infinity. Substituting each of these into (6.8), we obtain the unique solution $x = y = 2$.

Next, we consider the case of arbitrary n and show there are infinitely many solutions in positive rational numbers whenever there exists at least one solution in positive rational numbers. For this we need the following variation (see Silverman–Tate 1) of the Nagell–Lutz theorem.

Lemma 6.1. *Consider a non-singular elliptic curve*

$$y^2 = x^3 + ax^2 + bx$$

in Weierstrass form where a, b are integers. If (x, y) is a point of finite order with $y \neq 0$, then $x|b$ and

$$x + a + \frac{b}{x}$$

is a perfect square.

From the theory of torsion points on elliptic curves, we obtain the following theorem.

Theorem 6.4. *If $n \geq 6$, then the torsion points of $E_n : Y^2 = X^3 + (n^2 - 8)X^2 + 16X$ are the points*

$$(0, 0), (4, -4n), (4, 4n), \infty.$$

Proof. It is easy to confirm that $(0, 0)$, $(4, -4n)$, $(4, 4n)$, ∞ are torsion points of E_n. We would like to prove that there are no other torsion points of E_n. When $Y = 0$, note that $n \geq 6$ implies $X^3 + (n^2 - 8)X^2 + 16X = 0$ has the unique rational solution $X = 0$.

When $Y \neq 0$, suppose (X, Y) is a torsion point of E_n. From Lemma 6.1, it follows that $X|16$ and $X + n^2 - 8 + \frac{16}{X}$ is a perfect square. Since $X|16$, we must have $X \in \{1, 2, 4, 8, 16\}$.

If $X = 4$, then $X + n^2 - 8 + \frac{16}{X} = n^2$ is a perfect square. If $X \in \{1, 2, 8, 16\}$, then $X + n^2 - 8 + \frac{16}{X} = n^2 + 2$ or $n^2 + 9$, but $n^2 + 2$ is not a perfect square, and $n^2 + 9$ is a perfect square if and only if $n = 0, 4$.

It follows that when $Y \neq 0$, if (X, Y) is a torsion point of E_n, then $X = 4$ and therefore $Y = \pm 4n$. This completes the proof of Theorem 6.4. $\qquad\square$

Theorem 6.5. *If the abcd equation has any solutions in positive rational numbers for some $n \geq 6$, then it has infinitely many.*

Proof. It is necessary only to show that for any $n \geq 6$, if $E_n :$ $Y^2 = X^3 + (n^2 - 8)X^2 + 16X$ has a rational point $P_0(X_0, Y_0)$ satisfying $X_0 < 0$, then in fact there are infinitely many rational points (X, Y) satisfying $X < 0$. In fact, according to Theorem 6.4, any such $P_0(X_0, Y_0)$ is not a torsion point, so the points $[n]P_0$ are all distinct. We need only to prove that $[3]P_0$ satisfies $X < 0$, from which it follows inductively that every $[2k + 1]P_0$ satisfies $X < 0$. We prove this now.

First, if the lines $Y = kX + b$ where $b \neq 0$ and $Y^2 = X^3 + (n^2 - 8)X^2 + 16X$ intersect one another, we can substitute to obtain

$$X^3 + (n^2 - 8 - k^2)X^2 + (16 - 2kb)X - b^2 = 0$$

from which we see that the product of all roots of this polynomial is positive. Therefore, the intersection of E_n and the tangent to E_n through P_0 satisfy $X > 0$ (since there is a twice repeated root $X_0 < 0$), and by symmetry we find that $[2]P_0$ satisfies $X > 0$. Moreover, the intersection of E_n and the line passing through P_0 and $[2]P_0$ satisfy $X < 0$ (since $X_0 < 0$, and $[2]P_0$ satisfies $X > 0$). Therefore, by symmetry again we find that $[3]P_0$ satisfies $X < 0$. This completes the proof of Theorem 6.5. □

Example 6.2. The equation

$$13 = x + \frac{1}{x} + y + \frac{1}{y} \tag{6.9}$$

has infinitely many solutions in positive rational numbers.

By Theorem 6.3, we need only to find solutions of the elliptic curve

$$E_{13} : Y^2 = X^3 + 161X^2 + 16X, \ X < 0.$$

With the Magma package, we find that the rank of E_{13} is 1, and, following Mordell's theorem, we have a generating point $P(X, Y) = (-100, 780)$. By the group law, it follows that every rational point of E_{13} with $X < 0$ is given by $[2k + 1]P$, where k is any non-negative

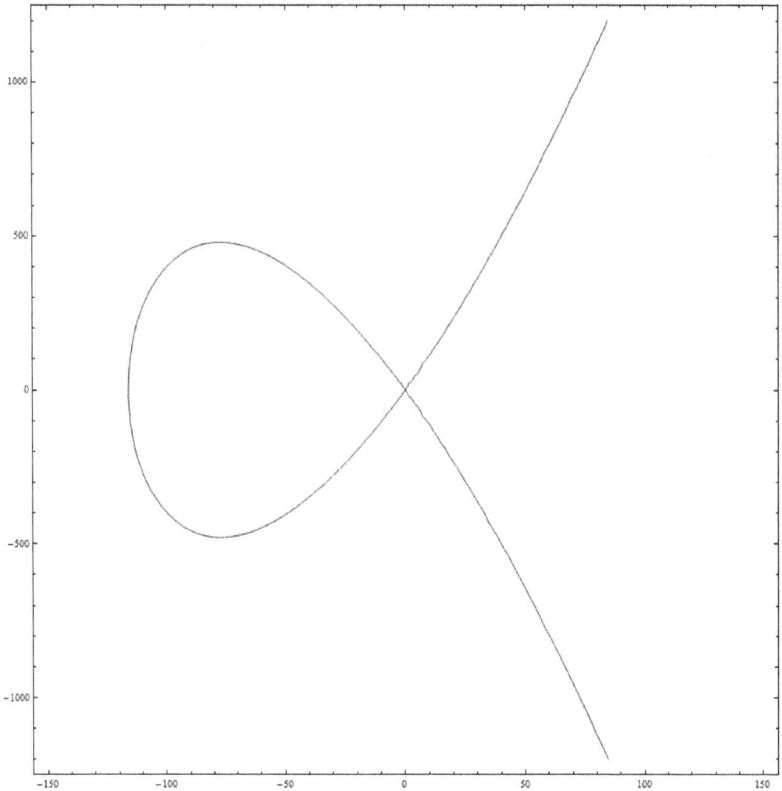

Figure 6.2.   Elliptic Curve E_{13}.

integer. Substituting into (6.7), we obtain

$$[1]P = (-100, 780)(k = 0), (x, y) = \left(\frac{2}{5}, 10\right);$$

$$[3]P = \left(-\frac{6604900}{776161}, \frac{71411669940}{683797841}\right)(k = 1),$$

$$(x, y) = \left(\frac{924169}{228730}, \frac{1347965}{156818}\right);$$

$$[5]P = \left(-\frac{31274879093702500}{57589364171021281}, \frac{85900073394621020231661900}{13820171278324441779434321}\right)$$

$$(x, y) = \left(\frac{33896240819350898}{3149745790659725}, \frac{12489591059767450}{8548281631402489}\right); \cdots,$$

from which we conclude that (6.9) has infinitely many solutions in rational numbers.

Finally, we still have the following question.

Question 6.4. Is there a better way than the above to determine whether or not the *abcd* equation has solutions for a given positive integer n.

Remarks. In comments to the author, Ye Tian has observed that by checking the ε divisors of the elliptic curve E_n, it can be deduced from the Birch and Swinnerton–Dyer conjecture (BSD) that if the number of distinct prime factors of a positive odd number n has the same parity as the number of distinct prime factors of $n^2 - 16$ which are congruent to 1 modulo 4, then there must exist a rational array (a, b, c, d) satisfying (6.1), although unfortunately it is not necessarily the case that all of these must be positive. Pan Jinzhao determined the solutions for $n = 11, 15$, and 19, respectively,

$$\left(-\frac{64}{9}, 1, -\frac{15}{8}, \frac{3}{40}\right), \left(-64, 1, -\frac{7}{24}, \frac{3}{56}\right), \left(-\frac{14161}{576}, 1, -\frac{1320}{1547}, \frac{312}{6545}\right).$$

6.3   An Ancient Problem

A congruent number is a number than can be obtained as the area of a right triangle with rational length sides. The simplest example is the number 6, which is the area of the right triangle with sides $(3, 4, 5)$. In fact, 6 was the first congruent number discovered by mathematicians, but it is not the smallest congruent number. It is clear from the definition and example that congruent numbers are entirely unrelated to the concept of congruence from elementary number theory.

Sometime before the year 972, the following question was posed in an Arabic manuscript: When can a positive integer n be the common difference of an arithmetic progression formed by three rational square numbers? In other words, under what conditions does there exist a rational number x such that

$$x - n, x, x + n$$

are all the squares of rational numbers?

A positive integer n satisfying this condition is precisely a congruent number. Indeed, suppose

$$x - n = a^2, x = b^2, x + n = c^2.$$

Then

$$2x = c^2 + a^2, 2n = c^2 - a^2 = (c - a)(c + a),$$

where both $c - a$ and $c + a$ are both rational numbers forming the lengths of two sides around the right angle of a right triangle, since

$$(c - a)^2 + (c + a)^2 = 2(c^2 + a^2) = (2x)^2.$$

So we have a right triangle with rational sides and area given by $n = (c - a)(c + a)/2$; therefore, n is a congruent number.

We can further transform the problem as follows. Let n be a positive integer. Then if there exists a rational number x such that

$$x^2 \pm n \tag{6.10}$$

is the square of a rational number, then n is a congruent number.

Suppose that (x, y, z) is a Pythagorean triple:

$$x^2 + y^2 = z^2. \tag{6.11}$$

Then we have

$$z^2 \pm 2xy = (x \pm y)^2.$$

It follows that $2xy$ is a congruent number; assuming that one of x, y is even and the other odd, then also $n = xy/2$ is a congruent number. It is straightforward to check that equations (6.10) and (6.11) establish a bijection between solutions.

The so-called congruent number problem is to produce some simple rule to determine whether or not any given natural number is a congruent number. It is clear that for any positive integer m, n, $m^2 n$ is a congruent number if and only if n is a congruent number. Therefore, it is sufficient to consider only square-free positive integers. Among the positive integers less than 6, then, we can exclude 4 and consider only the numbers 1, 2, 3, 5. Generating a few more examples from Pythagorean triples, the smallest such triple (3, 4, 5) shows that both 24 and 6 are congruent numbers; the triples (5, 12, 13), (8, 15, 17), (7, 24, 25) show that also 120 and 30, 240, 60, and 15, and 336, 84, and 21 are all congruent numbers.

It is believed that the problem of congruent numbers was studied by the Persian mathematician and engineer Al-Karaji (ca. 970–1029), who is best known for two mathematical works, one of which is an extremely rich treatise on algebra known as *Al-fakhri fi al-jabr wa al-muqabala*, written in 1010 and named in honor of his patron, the far-sighted ruler of Baghdad, where Al-Karaji lived.

Around 1220, Fibonacci proved that 5 is a congruent number, following the suggestion of a friend of his in Palermo, the capital of Sicily. In particular, he found the corresponding right triangle with side lengths $(\frac{3}{2}, \frac{20}{3}, \frac{41}{6})$ (see Figure 6.3).

Fibonacci, also known as Leonardo Pisano, was born in Pisa and traveled around North Africa with his tax collector father as a child. He was the first to introduce Arabic numerals to Europe and also proposed the Fibonacci sequence as a topic of interest in the form of a famous problem concerning the reproduction of rabbits. He wrote his most influential work, the *Liber Abaci* (*Book of Calculation*) in 1202, and included in it a substantial wealth of Greek, Egyptian, Arabic, Indian, and even Chinese mathematics. Fibonacci

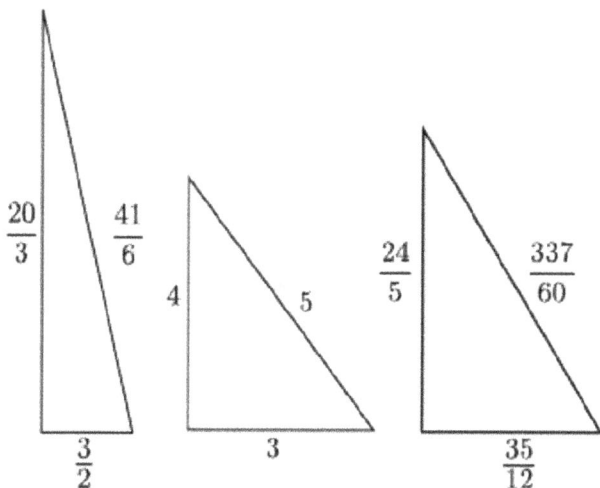

Figure 6.3. The three smallest congruent numbers.

was the most important mathematician of the European Middle Ages.

It is worth mentioning that we can easily prove that there is no right triangle with positive integer side lengths and area 5. Among the smallest congruent numbers 6, 15, 21, and 30 obtained from the Pythagorean triples above, 6 and 30 correspond to right triangles with positive integer side lengths, but for 15 and 21 there are no such corresponding triangles.

It is natural then to ask whether the congruent number 5 can be obtained from a Pythagorean triple, and this is precisely how Fibonacci obtained the rational side lengths of the right triangle corresponding to it. The 10th-century Arab mathematician Mohammed Ben Alhocain had discovered that for any numbers d, e, substituting $n = k = \frac{de(d+e)}{d-e}$ into (6.10) yields the identity

$$\left(\frac{d^2 + e^2}{2(d - e)}\right)^2 \pm k = \left(\frac{d + e}{2} \pm \frac{de}{d - e}\right)^2. \qquad (6.12)$$

With $d = 5$, $e = 4$, then $k = 180$, and substituting into (6.10) gives

$$x^2 + k = \left(\frac{49}{2}\right)^2, \quad x^2 - k = \left(\frac{31}{2}\right)^2,$$

where $x = \frac{41}{2}$, $c = \frac{49}{2}$, $a = \frac{31}{2}$. This produces the lengths of the two sides about the right angle as $c + a = 40$, $c - a = 9$. In other words, the congruent numbers 180, 45, 20, and 5 can all be obtained from the Pythagorean triple (9, 40, 41).

In the 18th century, Euler discovered that 7 is a congruent number. Putting $e = 112$, $d = 63$, $k = 252000$ in (6.12) and substituting, we get

$$x^2 + k = \left(\frac{463}{2}\right)^2, \quad x^2 - k = \left(\frac{113}{2}\right)^2,$$

where $x = \frac{337}{2}$, $c = \frac{463}{2}$, $a = \frac{113}{2}$, from which we obtain the side lengths $c + a = 288$ and $c - a = 175$. So the congruent numbers 25200, 6300, 2800, 1575, 1008, 700, 252, 175, 112, 63, 28, and 7 are all obtained from the Pythagorean triple (175, 288, 337).

The question then becomes as follows: Can every congruent number be obtained from a primitive Pythagorean triple? It is worth noting here that the correspondence between Pythagorean triples and congruent numbers is not injective; for example, the congruent number 210 can be obtained from both the Pythagorean triples (20, 21, 29) and (12, 35, 37) by setting $e = 14$, $d = 6$ or $e = 7$, $d = 5$ respectively in (6.12).

Fibonacci also discovered that if $a > b$ and any three of the four numbers a, b, $a - b$, $a + b$ are squares, then the fourth must be a congruent number. For example, with $a = 16$, $b = 9$, $a + b = 25$, so $a - b = 7$ is a congruent number. He claimed moreover to have discovered that 1 is not a congruent number, but his proof was flawed; four centuries later, Fermat gave a correct proof.

Theorem 6.6. *The number 1 is not a congruent number.*

This result also establishes the validity of Fermat's last theorem in the case $n = 4$, in other words that the equation $x^4 + y^4 = z^4$ has no solutions in positive integers, one of the few results for which Fermat provided proof during his lifetime. As we have mentioned above, it can be shown that all square numbers, including 4, are not congruent numbers. The latter conclusion is equivalent to the fact that the only rational solutions of the elliptic curve $y^2 = x^3 - x$ are $(x, y) = (0, 0)$ and $(\pm 1, 0)$. More generally, a number d is a congruent number if and only if the elliptic curve

$$y^2 = x^3 - d^2 x$$

has rational solutions other than $(x, y) = (0, 0)$ and $(\pm d, 0)$. This equivalence is verified by establishing a bijection between the two sets, using Lemmas 1.4 and 1.5 in *Number Theory I–II* (see Kato–Kurokawa–Saito 1).

We prove next that 2 and 3 are not congruent numbers.

Theorem 6.7. *The numbers 2 and 3 are not congruent numbers.*

Proof. Let $A^2 + B^2 = C^2$, $\frac{AB}{2} = N$ with A, B, C positive rational numbers with the lowest common multiple of their denominators given by s so that $a = sA$, $b = sB$, $c = sC$ are integers satisfying $a^2 + b^3 = c^2$, $\frac{ab}{2} = s^2 N$. It is clear that (a, b, c) is a primitive Pythagorean triple, so following Euclid's formula, we have $(a, b, c) = (m^2 - n^2, 2mn, m^2 + n^2)$ for some integers m, n, exactly one of which is even and the other odd, with $(m, n) = 1$. Then

$$\frac{ab}{2} = (m^2 - n^2)mn = (m + n)(m - n)mn = s^2 N.$$

From this, we use the method of infinite descent to complete the proof, starting with the case $N = 2$; without loss of generality, we suppose that n is even. Since $m + n$, $m - n$, m, n are relatively prime in pairs, we have

$$m + n = x^2, \quad m - n = y^2, \quad m = z^2, \quad n = 2w^2 \qquad (6.13)$$

where also x, y, z, w are relatively prime in pairs, and x, y, z are odd. From this we get that

$$2n = 4w^2 = (m + n) - (m - n) = x^2 - y^2 = (x + y)(x - y).$$

Since $(x + y, x - y) = (2x, x - y) = (2, x - y) = 2$, we find that

$$w^2 = \left(\frac{x + y}{2}\right)\left(\frac{x - y}{2}\right).$$

Let $\frac{x+y}{2} = u^2$, $\frac{x-y}{2} = v^2$ with $(u, v) = 1$. Since both x and y are odd, we must have that one of $x - y \equiv 0 \pmod 4$ or $x + y \equiv 0 \pmod 4$, say without loss of generality that $x - y \equiv 0 \pmod 4$, in which case u is odd and v is even, say $v = 2d$. Then

$$x = \frac{x + y}{2} + \frac{x - y}{2} = u^2 + v^2 = u^2 + 4d^2,$$

$$y = \frac{x + y}{2} - \frac{x - y}{2} = u^2 - v^2 = u^2 - 4d^2.$$

Substituting these identities into (6.13), we have

$$m = \frac{x^2 + y^2}{2} = u^4 + 16v^4 = z^2,$$

which gives the Pythagorean triple $(u^2, 4v^2, z)$ defining a triangle with area $2u^2 v^2$ corresponding to the congruent number 2, with hypotenuse

$$z \le m < m^2 + n^2 = c,$$

in other words generating a right triangle corresponding to the congruent number 2 with positive integer side lengths but a smaller hypotenuse than that with which we started; we conclude that 2 cannot be a congruent number.

Moving on to the second case, suppose that $N = 3$ is a congruent number. We have

$$(m + n)(m - n)mn = 3s^2.$$

Without loss of generality, we take m to be odd and n to be even and consider the four possible cases

(1) $m + n = 3x^2$, $m - n = y^2$, $m = z^2$, $n = (2w)^2$,
(2) $m + n = x^2$, $m - n = 3y^2$, $m = z^2$, $n = (2w)^2$,
(3) $m + n = x^2$, $m - n = y^2$, $m = 3z^2$, $n = (2w)^2$,
(4) $m + n = x^2$, $m - n = y^2$, $m = z^2$, $n = 3(2w)^2$,

where in each case x, y, z are odd numbers.

In the first case, we have

$$2n = 8w^2 = (m + n) - (m - n) = 3x^2 - y^2,$$

that is,

$$y^2 = 3x^2 - 8w^2,$$

which is easily seen to have no solutions in integers by considering both sides modulo 4.

In the second case,

$$2n = 8w^2 = (m + n) - (m - n) = x^2 - 3y^2$$

or

$$x^2 = 3y^2 + 8w^2.$$

Again we see that there are no solutions modulo 4 and therefore no integer solutions.

For the third case, we have

$$\frac{n}{2} = 2w^2 = \left(\frac{x + y}{2}\right)\left(\frac{x - y}{2}\right).$$

Here $\frac{x+y}{2} = u^2$, $\frac{x-y}{2} = v^2$ with $(u,v) = 1$, and one of u, v odd, the other even, say u is odd; writing $v = 2d$, we get

$$x = \frac{x+y}{2} + \frac{x-y}{2} = u^2 + v^2 = u^2 + 4d^2,$$

$$y = \frac{x+y}{2} - \frac{x-y}{2} = u^2 - v^2 = u^2 - 4d^2,$$

$$m = \frac{x^2 + y^2}{2} = u^4 + 16v^4 = 3z^2.$$

Reducing modulo 4, we see that there are no solutions in virtue of the fact that u and z are both odd.

Finally, in the fourth case,

$$\frac{n}{2} = 6w^2 = \left(\frac{x+y}{2}\right)\left(\frac{x-y}{2}\right).$$

Assuming $x - y \equiv 0 \pmod 4$, there are two possibilities. The first is

$$\frac{x+y}{2} = 3u^2, \quad \frac{x-y}{2} = 2v^2 \text{ with } (u,v) = 1,$$

which gives

$$x = \frac{x+y}{2} + \frac{x-y}{2} = 3u^2 + 2v^2,$$

$$y = \frac{x+y}{2} - \frac{x-y}{2} = 3u^2 - 2v^2,$$

and therefore

$$m = \frac{x^2 + y^2}{2} = 9u^4 + 4v^4 = z^2.$$

Then $(3u^2, 2v^2, z)$ is a primitive Pythagorean triple representing a triangle with area $3u^2v^2$ corresponding to the congruent number 3, with hypotenuse satisfying

$$z \le m < m^2 + n^2 = c.$$

So we obtain an infinite descent of triangles with positive integer side lengths ordered by hypotenuse length corresponding to the congruent number 3, showing that 3 cannot be a congruent number under this condition.

The second possibility is

$$\frac{x+y}{2} = u^2, \quad \frac{x-y}{2} = 6v^2 \quad \text{with } (u, v) = 1,$$

which gives

$$x = \frac{x+y}{2} + \frac{x-y}{2} = u^2 + 6v^2,$$

$$y = \frac{x+y}{2} - \frac{x-y}{2} = u^2 - 6v^2.$$

Therefore,

$$m = \frac{x^2 + y^2}{2} = u^4 + 36v^4 = z^2.$$

Then $(u, 6v^2, z)$ is a primitive Pythagorean triple representing a triangle with area $3u^2v^2$ corresponding to the congruent number 3 and with hypotenuse satisfying

$$z \le m < m^2 + n^2 = c.$$

Therefore, we conclude by the same argument above that 3 cannot be a congruent number under this condition and in fact under the assumption that $x - y \equiv 0 \pmod 4$; similarly, we can prove that 3 cannot be a congruent number under the assumption that $x + y \equiv 0 \pmod 4$. Therefore, 3 is not a congruent number. This completes the proof of Theorem 6.7. $\square$

We conclude from this that 5, 6, and 7 are the smallest, second smallest, and third smallest congruent numbers, respectively.

Mathematicians have verified that there are 23 congruent numbers among the first 50 positive integers, namely 5, 6, 7, 13, 14, 15, 20, 21, 22, 23, 24, 28, 29, 30, 31, 34, 37, 38, 39, 41, 45, 46, 47; among the squarefree numbers less than 1000, 361 are congruent numbers, and 247 are not. The congruent numbers also exhibit various properties and special cases; for example, the product $n^3 - n$ of three consecutive positive integers must be a congruent number, numbers of the form $4n^3 + n$, $n^4 - m^4$, $n^4 + 4m^4$, $2n^4 + 2m^2$, are necessarily congruent numbers, and so on.

In the second half of the 20th century, mathematicians discovered that the congruent number problem, like Fermat's last theorem, is closely related to elliptic curves; in fact, an integer n is a congruent number if and only if the system of equations

$$\begin{cases} a^2 + b^2 = c^2 \\ \frac{1}{2}ab = n \end{cases} \tag{6.14}$$

has a solution (a, b, c) in positive rational numbers.

Adding or subtracting four times the second equation from the first gives

$$(a \pm b)^2 = c^2 \pm 4n.$$

Then taking the product of the equations and dividing by 16,

$$\left(\frac{a^2 - b^2}{4}\right)^2 = \left(\frac{c}{2}\right)^4 - n^2.$$

This shows that if n is a congruent number, then the equation $u^4 - n^2 = v^2$ has the solution $u = \frac{c}{2}$, $v = \frac{a^2 - b^2}{4}$ in rational numbers. Multiplying both sides by u^2 and setting $x = u^2 = (\frac{c}{2})^2$, $y = uv = \frac{(a^2 - b^2)c}{8}$, we see that (x, y) is a rational point along the elliptic curve given by

$$y^2 = x^3 - n^2 x. \tag{6.15}$$

Conversely, if x and y satisfy (6.15) and $y \neq 0$, then it is easy to see that

$$a = \frac{x^2 - n^2}{y}, \quad b = \frac{2nx}{y}, \quad c = \frac{x^2 + n^2}{y}$$

satisfy (6.14). This establishes a bijective correspondence between solutions (a, b, c) and (x, y). From the theory of elliptic curves, it can be shown that the torsion points of (6.15) are the points that make $y = 0$ so that the existence of rational points with nonzero y is equivalent to the curve having positive rank.

We can further use the properties of elliptic curves to establish the following results: When $p \equiv 3 \pmod 8$ is prime, p is not a congruent number, but $2p$ is a congruent number; when $p \equiv 5 \pmod 8$ is prime, p is a congruent number, but $2p$ is not a congruent number; when $p \equiv 7 \pmod 8$ both p and $2p$ are congruent numbers.

The negative conclusions in the first two results just mentioned were obtained by L. Bastien (see Bastien 1) in 1915, and the positive conclusion in the second result by the German radio engineer Kurt Heegner (1893–1965) (see Heegner 1), who was also the first to prove that

there are infinitely many squarefree congruent numbers.

In 2014, Ye Tian proved (see Tian 1) that for any positive integer k, there are contained in each of the residue classes $n \equiv 5, 6, 7 \pmod 8$ infinitely many square-free congruent numbers with k prime factors. Ye Tian, Yuan Xinyi, and Zhang Shouwu also obtained (see Tian–Yuan–Zhang 1) various necessary and sufficient conditions for a number n to be a congruent number and believed that these results implied that the set of congruent numbers has positive density. Later, Alexander Smith announced a proof (see Smith 1) that the density of congruent numbers in the residue classes $n \equiv 5$ or $7 \pmod 8$, the congruent numbers have a density of at least 62.9%, and in the residue class $n \equiv 6 \pmod 8$, the congruent numbers have a density of at least 41.9%.

According to the Goldfeld conjecture, congruent numbers and non-congruent numbers each account for half of all integers; more precisely, almost all numbers congruent to 5, 6, or 7 modulo 8 are congruent numbers, and almost all numbers congruent to 1, 2, or 3 modulo 8 are not. It can be proved that one of the consequences of the BSD conjecture is that every $n \equiv 5, 6, 7 \pmod 8$ is a congruent number.

In fact, the BSD conjecture states the following:

$L(E, s)$ *has Taylor expansion at* $s = 1$ *given by* $L(E, S) = c(s - 1)^m +$ (higher order terms) *where* $L(E, s)$ *is the Dirichlet L-function (Euler product) of an elliptic curve* E, $c \neq 0$ *is a constant,* $m = r = \operatorname{rank}(E(Q))$ *is the rank of* $E(Q)$, *the set of rational points of the elliptic curve.*

Poincaré had noticed that $E(Q)$ admits a group structure, and in 1922 Mordell proved that $E(Q)$ is a finitely generated abelian group, namely

$$E_n(Q) \cong T \bigoplus Z^{\operatorname{rank}(E_n)},$$

where T is the torsion subgroup, a special case of which we have already made use of in Example 6.3; in 1977, J. Mazur proved that $|T| \leq 16$.

It is possible to prove by elementary methods that a number n is a congruent number if and only if $E(Q)$ is infinite, in other words when its rank $r > 0$. Then assuming the BSD conjecture, we can obtain a necessary and sufficient condition for n to be a congruent number, specifically $s = 1$ is a zero of $L(E_n, s)$. This necessary and sufficient condition is also referred to as the weak BSD conjecture (see Wang, 2004).

On the other hand, let

$$\Lambda(s) = \left(\frac{\sqrt{N}}{2\pi} \right)^s \Gamma(s) L(E_n, S),$$

where $\Gamma(s)$ is the Gamma function and

$$N = \begin{cases} 32n^2 & \text{if } n \text{ is odd,} \\ 16n^2 & \text{if } n \text{ is even.} \end{cases}$$

Then $\Lambda(s)$ satisfies the functional equation

$$\Lambda(s) = \varepsilon \Lambda(2 - s),$$

where ε is the root number defined by

$$\varepsilon = \begin{cases} \left(\frac{-2}{n} \right) & \text{if } n \text{ is odd,} \\ \left(\frac{1}{n/2} \right) & \text{if } n \text{ is even} \end{cases}$$

in which ( ) is the Jacobi symbol.

Using a fast-converging series expression for $L(E_n, 1)$, we can obtain a sufficient condition for n to be a congruent number under the weak BSD conjecture, specifically that the root number of $L(E_n, 1)$ is $\varepsilon = -1$.

When $n \equiv 5 \pmod 8$,

$$\left(\frac{-2}{n}\right) = \left(\frac{2}{n}\right)\left(\frac{-1}{n}\right) = (-1)(+1) = -1.$$

When $n \equiv 6 \pmod 8$,

$$\left(\frac{-1}{n/2}\right) = (-1)^{\frac{n}{2}-1}{2} = -1.$$

When $n \equiv 7 \pmod 8$,

$$\left(\frac{-2}{n}\right) = \left(\frac{2}{n}\right)\left(\frac{-1}{n}\right) = (+1)(-1) = -1.$$

Therefore, it follows that every $n \equiv 5$, 6, or 7 (modulo 8) is a congruent number. Among congruent numbers not of this form, the smallest is 34, congruent to 2 modulo 8, corresponding to the triangle with side lengths $(225/30, 272/30, 353/30)$. The smallest congruent numbers congruent to 1 and 3 modulo 8 are 41, corresponding to $(40/3, 123/20, 881/60)$, and 219, corresponding to $(55/4, 1752/55, 7633/220)$, respectively.

In 1983, the American mathematician Jerrold B. Tunnell proved (see Tunnell 1) the following theorem.

Tunnell's Theorem. If n is an odd congruent number, then

$$\#\{n = 2x^2 + y^2 + 8z^2\}\# = 2\#\{n = 2x^2 + y^2 + 32z^2\}\#.$$

If n is an even congruent number, then

$$\#\{n = 2x^2 + y^2 + 16z^2\}\# = 2\#\{n = 2x^2 + y^2 + 64z^2\}\#.$$

Here $\#\{\ \}\#$ indicates the number of integer solutions to the equation within the brackets. Under the BSD conjecture, the converse of this theorem is also valid.

The criterion given by Tunnell's theorem is a practical one that can be used in conjunction with the method of exhaustion. For example, when $n = 1$, the number of solutions to the first equation is 2, namely $(0, \pm 1, 0)$, from which it follows that 1 is not a congruent number. Building upon this, research into the congruent number problem achieved a breakthrough.

It is worth mentioning that in 2013, two American mathematicians Jones and Rouse showed that, under the BSD conjecture, the question of whether or not the Fermat equation $x^3 + y^3 = z^3$ has non-trivial solutions over quadratic fields can be reduced to whether or not the numbers of integer solutions to two quadratic equations in three variables are equal.

In 2022, Qin Hourong proved (see Qin 1) that if n is an odd congruent number, then

$$\#\{n = x^2 + 2y^2 + 32z^2\}\# = \#\{n = 2x^2 + 4y^2 + 9z^2 - 4yz\}\#,$$

and if n is an even congruent number, then

$$\#\left\{\frac{n}{2} = x^2 + 4y^2 + 32z^2\right\}\# = \#\left\{\frac{n}{2} = 4x^2 + 4y^2 + 9z^2 - 4yz\right\}\#.$$

Under the BSD conjecture, the converse of this result also holds.

Even after determining that some given n is a congruent number, it is not easy in general to identify a right triangle with rational side lengths and area n. The German-born American mathematician Don B. Zagier (1951–) once calculated the triangle corresponding to $n = 157$, for which the numerator and denominator of the length of the hypotenuse had 47 and 45 digits, respectively.

6.4 New Congruent Numbers

As with other problems in number theory, congruent numbers have also been subject to generalizations, such as the t-congruent numbers and θ-congruent numbers, which we define now. Let n be a positive integer and t a positive rational number; then n is called a t-congruent number if there exists a positive triple (a, b, c) of rational numbers satisfying

$$a^2 = b^2 + c^2 - 2bc\frac{t^2 - 1}{t^2 + 1}, \quad bc\frac{2t}{t^2 + 1} = 2n.$$

When $t = 1$, this describes a classical congruent number. It is not hard to see that the conditions above are in correspondence with the elliptic curves

$$E_{n,t} : y^2 = x\left(x - \frac{n}{t}\right)(x + nt).$$

Next, let $0 < \theta < \pi$ be a real number, with $\cos\theta = \frac{s}{r}$ in reduced form. If $n\sqrt{r^2 - s^2}$ is the area of a triangle with rational length sides and interior angle θ, then n is called a θ-congruent number. When $\theta = \frac{\pi}{2}$, this again describes a classical congruent number.

In October 2012, in light of the fact that a triangle can be viewed as a trapezoid with one base length equal to zero, the author undertook to consider a new class of congruent numbers.

Definition 6.2. A positive integer n is called a congruent integer if it is the area of a right trapezoid as shown in the figure (Figure 6.4), where a, b, c are positive integers, d a non-negative integer, and $(b, c) = 1$.

Definition 6.3. A positive integer n is called a k-congruent number if it is the area of a right trapezoid as shown in the figure (Figure 6.4) with a, b, c, d positive rational numbers, $k \geq 2$ an integer, and $a = kd$.

According to Definition 6.3, we have in this case that

$$n = (a + d)b/2, (a - d)^2 + b^2 = c^2. \tag{6.16}$$

Definition 6.4. A positive integer n is called a d-congruent number if it is the area of a right trapezoid as shown in the figure (Figure 6.4) where a, b, c are positive rational numbers and d is a non-negative integer.

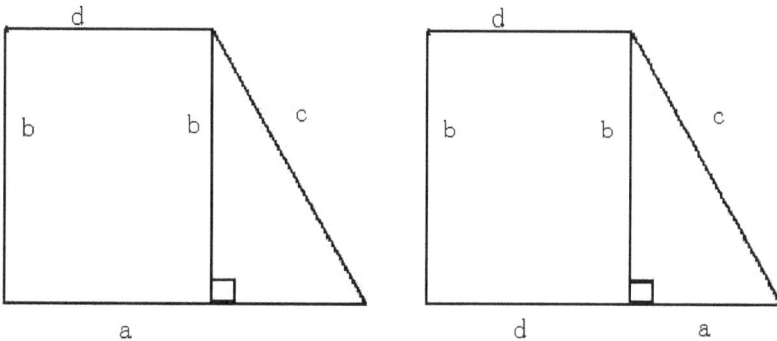

Figure 6.4. Congruent numbers.

According to Definition 6.4, we have in this case that

$$n = (a + 2d)b/2, \quad a^2 + b^2 = c^2. \tag{6.17}$$

We (Cai Tianxin and Zhang Yong, see Cai-Zhang 4) have investigated the three types of congruent numbers defined above one by one in turn.

According to Definition 6.2,

$$n = (a + d)b/2, \quad (a - d)^2 + b^2 = c^2, \quad (b, c) = 1.$$

Then from the properties of Pythagorean triples, we must have $(a - d, b) = 1$, where $a - b$ and d have different parities, and

$$(a - d, b) = (2xy, x^2 - y^2) \text{ or } (x^2 - y^2, 2xy),$$

where $x > y$, $(x, y) = 1$, and x and y have different parities. We conclude that

a positive integer n is a congruent integer if and only if $n = pk$
for some odd prime p with $k \geq \frac{p^2-1}{4}$, or $n = 2^i k$, $k \geq 2^{2i} - 1$,
$i \geq 1$ where k is any odd number. Conversely, $n > 1$ is not a
congruent integer if and only if n has one of the following forms

$$p, \quad p^2(p \neq 3), \quad pq\left(5 < p < q < \frac{p^2 - 1}{4}\right),$$

$$2^i(i \geq 0), \quad 2^i p(i \geq 2, \quad 2^{1+i/2} < p < 2^{2i} - 1),$$

where p and q are both primes.

Using the pigeonhole principle, we can prove that almost all positive integers are congruent integers. Furthermore, using analytic methods and the prime number theorem, we have the following theorem.

Theorem 6.8. *The $f(x)$ indicate the number of positive integers not exceeding x that are not congruent integers; then*

$$f(x) \sim \frac{cx}{\log x},$$

where $c = 1 + \ln 2$.

Remarks. In the case of classical congruent numbers, we can also separate out the congruent numbers as positive integers representing the area of a right triangle with positive integer side lengths; if the number of such congruent integers not exceeding x is $g(x)$, we do not have such an estimate, but it is possible to prove that

$$\frac{\sqrt{x}}{2} + O(1) < g(x) \le \frac{1}{2\sqrt[3]{4}} x^{\frac{2}{3}} + O(x^{\frac{5}{9}}).$$

Moving on, we use the theory of elliptic curves to study the k-congruent and d-congruent numbers.

Theorem 6.9. *Every positive integer n is k-congruent.*

Proof. Set

$$b = \left| \frac{x^2 - (k^2 - 1)^2 n^2}{(k+1)y} \right|, \quad d = \left| \frac{2nx}{y} \right|$$

in (6.16). We get a family

$$E_{n,k} : y^2 = x^3 - (k^2 - 1)n^2 x$$

of elliptic curves, with

$$a = \left| \frac{2knx}{y} \right|, \quad c = \left| \frac{x^2 + (k^2 - 1)^2 n^2}{(k+1)y} \right|.$$

When $k \ge 2$, $E_{n,k}$ is a special class of congruent number elliptic curves, which we refer to as k-congruent curves. Noting that $n^3 - n$ is a congruent number, if we set

$$p = n + 1, \quad q = n - 1,$$

then

$$4(n^3 - n) = pq(p^2 - q^2)$$

is a congruent number (according to Euclid's formula). Then if we put $k = n$, it follows that $E_{n,n}$ has positive rank, from which it follows that $n > 1$ is k-congruent.

When $n = 1$, take any $k = k_1^2$, $k_1 > 1$; then $E_{1,k} : y^2 = x^3 - (k_1^4 - 1)^2 x$, and since $k_1^4 - 1$ is a congruent number, this elliptic curve has infinitely many solutions, which shows that 1 is k-congruent. This completes the proof of Theorem 6.8. $\qquad\square$

Example 6.3. With $n = 1$, $k = 4$, we have

$$(a, b, c, d) = \left(2, \frac{4}{5}, \frac{17}{10}, \frac{1}{2}\right), \left(\frac{16}{15}, \frac{3}{2}, \frac{17}{10}, \frac{4}{15}\right),$$

$$\left(\frac{544}{161}, \frac{161}{340}, \frac{141121}{54740}, \frac{136}{161}\right).$$

With $k = n = 2$, from the elliptic curve $E_{2,2}$ we get

$$(a, b, c, d) = \left(\frac{8}{3}, 1, \frac{5}{3}, \frac{4}{3}\right), \left(\frac{80}{7}, \frac{7}{30}, \frac{1201}{210}, \frac{40}{7}\right),$$

$$\left(\frac{6808}{4653}, \frac{1551}{851}, \frac{7776485}{3959703}, \frac{3404}{4653}\right).$$

With $k = n = 3$, from the elliptic curve $E_{3,3}$ we get

$$(a, b, c, d) = \left(\frac{9}{4}, 2, \frac{5}{2}, \frac{3}{4}\right), \left(\frac{21}{40}, \frac{60}{7}, \frac{1201}{140}, \frac{7}{40}\right),$$

$$\left(\frac{851}{517}, \frac{4653}{1702}, \frac{7776485}{2639802}, \frac{851}{1551}\right).$$

Theorem 6.10. *Every positive integer is d-congruent.*

Proof. Set

$$\begin{cases} a = \frac{(3x - d^2 - 3n)(3x - d^2 + 3n)}{3(-3y + 3dx - d^3)} \\ b = \frac{2n(3x - d^2)}{-3y + 3dx - d^3} \\ c = \frac{(9 - 6d^2)x^2 + 9n^2 + d^4}{3(-3y + 3dx - d^3)} \end{cases}$$

in (6.17). We obtain a family

$$E_{n,d} : y^2 = x^3 - \frac{3n^2 + d^4}{3}x + \frac{(9n^2 + 2d^4)d^2}{27}$$

of elliptic curves, which we refer to as d-congruent curves. Putting $d = 3n$, we have

$$E_{n,3n} : y^2 = x^3 - (1 + 27n^2)n^2 x + 3n^4(1 + 18n^2),$$

with discriminant $\Delta = (4 + 81n^2)n^6 > 0$, which shows that $E_{n,3n}$ has no singularities.

We would like to show that for any positive integer n, the curve $E_{n,3n}$ contains infinitely many rational points, from which we can choose a point generating a solution (a, b, c, d) to (6.17). Note that the point $P(-6n^2, 3n^2)$ lies on the curve $E_{n,3n}$; then using the group law, we calculate

$$[2]P = \left(\frac{(27n^2 + 1)(243n^2 + 1)}{36}, -\frac{(81n^2 + 1)(6561n^4 + 324n^2 - 1)}{216} \right).$$

It is easy to see that for any $n \geq 1$, the x coordinate of $[2]P$ is not an integer. Then according to the Nagell–Lutz theorem, when $n \geq 4$, $[2]P$ has infinite order so that $E_{n,3n}$ has infinitely many rational points. Moreover, the point $[2]P$ gives rise to a, b, c, $d(= 3n)$ satisfying (6.17), specifically

$$\begin{cases} a = \frac{(729n^3 - 81n^2 + 27n + 1)(9n - 1)}{6(1 + 81n^2)}, \\ b = \frac{12n(1 + 81n^2)}{(1 + 9n)(729n^3 + 81n^2 + 27n - 1)}, \\ c = \frac{43046721n^8 + 2125764n^6 + 39366n^4 + 1620n^2 + 1}{6(1 + 81n^2)(1 + 9n)(729n^3 + 81n^2 + 27n - 1)}. \end{cases}$$

This completes the proof of Theorem 6.9. $\qquad\square$

Example 6.4. Considering $n = 1, 2, 3$ in turn we obtain from the elliptic curves $E_{n,3n}$ the following solutions:

$$(a, b, c, d) = \left(\frac{1352}{123}, \frac{123}{1045}, \frac{1412921}{128535}, 3 \right),$$

$$(a, b, c, d) = \left(\frac{94571}{1950}, \frac{7800}{117971}, \frac{11156645809}{230043450}, 6 \right),$$

$$(a, b, c, d) = \left(\frac{123734}{1095}, \frac{3285}{71722}, \frac{8874450677}{78535590}, 9 \right).$$

Working in the opposite direction, we have the following result: For any positive integer $k \geq 2$, there are infinitely many k-congruent positive integers n. According to the proof of Theorem 6.9, fixing $n = 1$, we have seen that 1 is k-congruent for infinitely many positive integers k; for $n \geq 2$, this is difficult to prove in general, but we have the following special case: If n is a perfect square, then n is k-congruent for infinitely many positive integers k. With the aid of numerical data obtained by computer search, we make the following conjecture.

Conjecture 6.3. *Every positive integer n is k-congruent for infinitely many positive integers k.*

We also have the following results for d-congruent numbers: For any positive integer d, a positive integer n is d-congruent if and only if $n \neq d^2$.

Chapter 7

Additive and Multiplicative Congruences and Other Topics

> A mathematician who is not also something of a poet will never be a complete mathematician.
>
> —*Karl Weierstrass*

7.1 Additive and Multiplicative Congruences

We have so far discussed a wide variety of additive and multiplicative equations; in this section we turn from identities to congruences, leading to some different conclusions. Both this section and the next were prepared in collaboration with Shen Zhongyan and Yang Peng (see Cai *et al.*, to appear). Our subject is the following pair of additive and multiplicative congruences:

$$n \equiv a + b \equiv ab \,(\mathrm{mod}\, p) \tag{7.1}$$

and

$$n \equiv a - b \equiv ab \,(\mathrm{mod}\, p), \tag{7.2}$$

where n is required to be a unit modulo p.

It is not difficult to prove that when $n = 1$, (7.1) has solutions if and only if $\left(\frac{-3}{p}\right) = 1$, that is, if $p = x^2 + 3y^2$, and (7.2) has solutions if and only if $\left(\frac{5}{p}\right) = 1$, that is if $p = 5x^2 - y^2$. Similarly when $n = 2$, (7.1) has solutions if and only if $\left(\frac{-1}{p}\right) = 1$, that is, if, $p = x^2 + y^2$, and (7.2) has solutions if and only if $\left(\frac{3}{p}\right) = 1$, that is, if $p = 3x^2 - y^2$.

For example,

$$1 \equiv 3 + 5 \equiv 3 \cdot 5 (\mathrm{mod}\, 7),$$

$$1 \equiv 5 - 4 \equiv 5 \cdot 4 (\mathrm{mod}\, 19),$$

$$2 \equiv 6 - 4 \equiv 6 \cdot 4 (\mathrm{mod}\, 11),$$

$$2 \equiv 6 + 9 \equiv 6 \cdot 9 (\mathrm{mod}\, 13).$$

Hereafter, when we say that n is a solution modulo p to (7.1) or (7.2), we mean that there exists a pair (a, b) satisfying (7.1) or (7.2) respectively. We now proceed to study the solutions of (7.1). It is easy to see that $n = 4$ is a solution to (7.1) modulo all p with $(a, b) = (2, 2)$.

Theorem 7.1. *For any odd prime p, there are exactly $\frac{p-1}{2}$ distinct n (modulo p) such that (7.1) has solutions, and, whenever there exists solution (a, b), it is unique apart from interchanging a and b.*

Proof. For $0 < a \neq 1 \leq p - 1$, the congruence

$$a + x - ax \equiv 0 (\mathrm{mod}\, p)$$

has exactlyone solution, $x \equiv \frac{a}{a-1} (\mathrm{mod}\, p)$ (when $a = 1$, there is no solution); we have the solution $x \equiv 2 (\mathrm{mod}\, p)$ for $a = 2$. It follows that congruence (7.1) has $\frac{p-3}{2} + 1 = \frac{p-1}{2}$ solutions. $\qquad\square$

For fixed n, suppose $(a_1, b_1), (a_2, b_2)$ are two solutions to (7.1); then

$$a_1 + \frac{a_1}{a_1 - 1} \equiv a_2 + \frac{a_2}{a_2 - 1} (\mathrm{mod}\, p),$$

or

$$\frac{(a_1 + a_2 - a_1 a_2)(a_1 - a_2)}{a_1 - 1} \equiv 0 (\mathrm{mod}\, p).$$

It follows that either $a_1 \equiv a_2 (\mathrm{mod}\, p)$, or $a_1 \equiv \frac{a_2}{a_2 - 1} (\mathrm{mod}\, p)$, $a_2 \equiv \frac{a_1}{a_1 - 1} (\mathrm{mod}\, p)$. From the former we determine that $b_1 \equiv b_2 (\mathrm{mod}\, p)$

and from the latter that $a_1 = b_2, a_2 = b_1$. This completes the proof of Theorem 7.1.

It is worth remembering here that the quadratic residues and quadratic non-residues, respectively also account for half of the units modulo p, that is, there are $\frac{p-1}{2}$ of each. The proof of this result makes use of Fermat's little theorem, which we did not need to prove Theorem 7.1; the proof of Theorem 7.3, concerning powers of solutions to (7.1), will make use of it.

In the following theorem, we set

$$S_+ = \{n \mid n \equiv a + b \equiv ab(\bmod p)\}$$

equal to the set of all n solving (7.1) modulo p.

Theorem 7.2. *Fix p an odd prime; then the product of all solutions to (7.1) satisfies the following congruence:*

$$\prod_{n_i \in S_+} n_i \equiv -2(\bmod p).$$

Proof. From Theorem 7.1, we see that (7.1) has the unique solution $(2,2)$ only when $n = 4$, and for other n, a and b in any solution belong to distinct residue classes modulo p. Therefore, apart from 1, which does not occur, and 2, which appears twice, a and b range across every other non-zero residue class modulo p. Then by Wilson's theorem, we have

$$\prod_{n_i \in S_+} n_i \equiv 2 \prod_{j=2}^{p-1} j \equiv -2(\bmod p).$$

This completes the proof of 7.2. □

Theorem 7.3. *Fix p an odd prime, $k \equiv s(\bmod p-1), 0 \leq s < p-1$; then the sum of the kth powers of the solutions to (7.1) satisfies*

$$\sum_{n_i \in S_+} n_i^k \equiv \begin{cases} 2^{2s-1} - \frac{C_{2s}^2}{2}(\bmod p) & \text{if} \quad s \neq 0 \\ \frac{p-1}{2}(\bmod p) & \text{if} \quad s = 0. \end{cases}$$

In order to prove Theorem 7.3, we need the following lemma (see Murty, 2009).

Lemma 7.1. *For any integer k and prime p, we have*

$$\sum_{x=1}^{p-1} x^k \equiv \begin{cases} 0 \pmod{p} & p-1 \nmid k, \\ -1 \pmod{p} & p-1 \mid k. \end{cases}$$

Proof of Theorem 7.3. For $0 < a_i \neq 1 \leq p-1$, (7.1) has solutions $x_i \equiv a_i/(a_i - 1) \pmod{p}$. From Theorem 7.1, it is clear that apart from $n = 2 \times 2 = 4$, the remaining solutions (a, b) appear in pairs, each with $a \neq b$. Therefore, for $p > 3$, we have

$$\sum n_i^k \equiv \frac{1}{2}\left(\sum_{j=2}^{p-1} \frac{j^{2k}}{(j-1)^k} - 2^{2k} \right) + 2^{2k}$$

$$\equiv \frac{1}{2} \sum_{j=2}^{p-1} \frac{(j-1+1)^{2k}}{(j-1)^k} + 2^{2k-1}$$

$$\equiv \frac{1}{2} \sum_{j=2}^{p-1} \sum_{t=0}^{2k} C_{2k}^t (j-1)^{t-k} + 2^{2k-1}$$

$$\equiv \frac{1}{2} \sum_{t=0}^{2k} C_{2k}^t \sum_{j=2}^{p-1} (j-1)^{t-k} + 2^{2k-1}$$

$$\equiv \frac{1}{2} \sum_{t=0}^{2k} C_{2k}^t \left(\sum_{j=1}^{p-1} j^{t-k} - (p-1)^{t-k} \right) + 2^{2k-1}$$

$$\equiv \frac{1}{2} \sum_{t=0}^{2k} C_{2k}^t \left(\sum_{j=1}^{p-1} j^{t-k} - (-1)^{t-k} \right) + 2^{2k-1}.$$

For $k < p-1$, Lemma 7.1 shows that

$$\sum n_i^k \equiv 2^{2k-1} - \frac{C_{2k}^k}{2} \pmod{p}.$$

In general, if $p - 1 \nmid k$ and $k \equiv s \pmod{p-1}$,

$$\sum n_i^k \equiv 2^{2s-1} - \frac{C_{2s}^s}{2} \pmod{p}.$$

If $p - 1 | k$, in other words if $s = 0$, then Fermat's little theorem gives

$$\sum n_i^k \equiv \sum 1 \equiv \frac{p-1}{2} \pmod{p},$$

Finally, it is easy to check directly that the result holds when $p = 3$, completing the proof.

We also obtain the following corollary from Fermat's little theorem.

Corollary 7.1. *For primes p,*

$$\sum_{n \in S+} \frac{1}{n} \equiv \frac{1}{8} \pmod{p}, \ \text{provided } p > 3,$$

$$\sum_{n \in S_+} \frac{1}{n^2} \equiv \frac{1}{32} \pmod{p}, \ \text{provided } p > 5.$$

Let R and N denote, respectively, the set of quadratic residues and the set of quadratic nonresidues modulo p; we define the following sets:

$$RR = \left\{ a \in Z_p^* : a \in R, a + 1 \in R \right\}$$
$$RN = \left\{ a \in Z_p^* : a \in R, a + 1 \in N \right\}$$
$$NR = \left\{ a \in Z_p^* : a \in N, a + 1 \in R \right\}$$
$$NN = \left\{ a \in Z_p^* : a \in N, a + 1 \in N \right\}.$$

Then we have (see Cai, 2021):

$$|RR| = \frac{p - 4 - \left(\frac{-1}{p}\right)}{4}, |RN| = \frac{p - \left(\frac{-1}{p}\right)}{4},$$

$$|NR| = |NN| = \frac{p - 2 - \left(\frac{-1}{p}\right)}{4}.$$

Inspired by this result, we investigate the intersections of S_+ with R and N, and obtain the following theorem.

Theorem 7.4. *Let p be an odd prime. Then*

$$|S_+ \cap R| = \frac{1}{4}\left(p - \left(\frac{-1}{p}\right)\right)$$

$$|S_+ \cap N| = \frac{1}{4}\left(p - 2 + \left(\frac{-1}{p}\right)\right).$$

Proof. Fix $n \in S_+$, say $n \equiv a + b \equiv ab \pmod{p}$; then $(a-1)(b-1) \equiv 1 \pmod{p}$. Consider the case $a - 1 \equiv b - 1 \pmod{p}$; if $a - 1 \equiv b - 1 \equiv 1 \pmod{p}$, then $a \equiv b \equiv 2 \pmod{p}, n \equiv ab \equiv 4 \pmod{p}, n \in S_+$. If $a - 1 \equiv b - 1 \equiv -1 \pmod{p}, a \equiv b \equiv 0 \pmod{p}$, $n \equiv ab \equiv 0 \pmod{p}$ is not a unit modulo p, in particular n does not belong to S_+.

Then suppose $n \in S_+ \cap R$, so that $\left(\frac{n}{p}\right) = 1$. Noting that $n \equiv ab \equiv \frac{a^2}{a-1} \pmod{p}$, we see that also $\left(\frac{a-1}{p}\right) = 1$, and similarly that $\left(\frac{b-1}{p}\right) = 1$. Then $|S_+ \cap R|$ is given by pairs $(a - 1, b - 1)$ modulo p satisfying $(a-1)(b-1) \equiv 1 \pmod{p}$ and $\left(\frac{a-1}{p}\right) = \left(\frac{b-1}{p}\right) = 1$. Apart from $a - 1 \equiv b - 1 \equiv \pm 1 \pmod{p}$, the remaining such pairs consist of two distinct residues modulo p.

If $p \equiv 1 \pmod 4$, then ± 1 are quadratic residues modulo p, so

$$|S_+ \cap R| = \frac{\frac{p-1}{2} - 2}{2} + 1 = \frac{p-1}{4} = \frac{1}{4}\left(p - \left(\frac{-1}{p}\right)\right).$$

If $p \equiv 3 \pmod 4$, then 1 is a quadratic residue modulo p but -1 is not, so

$$|S_+ \cap R| = \frac{\frac{p-1}{2} - 1}{2} + 1 = \frac{p+1}{4} = \frac{1}{4}\left(p - \left(\frac{-1}{p}\right)\right).$$

Finally, counting the remaining elements,

$$|S_+ \cap N| = |S_+| - |S_+ \cap R| = \frac{1}{4}\left(p - 2 + \left(\frac{-1}{p}\right)\right).$$

This completes the proof of Theorem 7.4. $\qquad\qquad \square$

Theorem 7.5. *For $p > 3$ prime and an integer $k \equiv s \,(\mathrm{mod}\, p - 1)$ where $0 \le s < p - 1$,*

$$\sum_{n \in S_+ \cap R} n^k \equiv \begin{cases} -\frac{1}{4}\left(\frac{-1}{p}\right) \,(\mathrm{mod}\, p) & \text{if } s = 0 \\ 2^{2s-1} - \frac{1}{4}\binom{2s}{2} \,(\mathrm{mod}\, p) & \text{if } 0 < 2 < \frac{p-1}{2} \\ 2^{2s-1} - \frac{1}{4}\left(\binom{2s}{s} + 2\binom{2s}{s-\frac{p-1}{2}}\right) & \text{if } \frac{p-1}{2} \le s < p - 1 \\ (\mathrm{mod}\, p) \end{cases}$$

We need the following lemmas in order to prove Theorem 7.5.

Lemma 7.2 (see Mordell, 1961). For any odd prime p, we have

$$\left\{\left(\frac{p-1}{2}\right)!\right\}^2 \equiv (-1)^{\frac{p+1}{2}} \,(\mathrm{mod}\, p).$$

Lemma 7.3. *For any odd prime p and positive integer l, we have*

$$\sum_{a \in R} a^l \equiv \begin{cases} 0 \,(\mathrm{mod}\, p) & \text{if } \frac{p-1}{2} \nmid l \\ \frac{p-1}{2} \,(\mathrm{mod}\, p) & \text{if } \frac{p-1}{2} \mid l \end{cases}$$

in particular, when $p = 3$,

$$\sum_{a \in R} a^l \equiv 1 (\mathrm{mod}\, 3).$$

Proof. It is easy to check that the result holds when $p = 3$ or $\frac{p-1}{2} \mid l$. If $p > 3$ and $\frac{p-1}{2} \nmid l$, then

$$\sum_{a \in R} a^l \equiv \sum_{i=1}^{\frac{p-1}{2}} i^{2l} \equiv \frac{1}{2}\sum_{i=1}^{p-1} i^{2l} \equiv 0(\mathrm{mod}\, p).$$

$\square$

Proof of Theorem 7.5. If $s = 0$, the claim follows directly from Theorem 7.3. Otherwise, for $1 < a \le p - 1$,

$$b \equiv \frac{a}{a-1}(\mathrm{mod}\, p)$$

Apart from $n = 4$, any solution (a, b) to (71) must have $a \not\equiv b \pmod{p}$. Then for $p > 3, 0 < s < p - 1$,

$$\sum_{n \in S_+ \cap R} n^k \equiv \sum_{n \in S_+ \cap R} n^s \equiv \frac{1}{2} \left(\sum_{a-1 \in R} \frac{a^{2s}}{(a-1)^s} - 2^{2s} \right) + 2^{2s} 2^{2s}$$

$$\equiv \frac{1}{2} \sum_{a-1 \in R} \frac{(a-1+1)^{2s}}{(a-1)^s} + 2^{2s-1}$$

$$\equiv \frac{1}{2} \sum_{a-1 \in R} \sum_{t=0}^{2s} \binom{2s}{t} (a-1)^{t-s} + 2^{2s-1}$$

$$\equiv \frac{1}{2} \sum_{t=0}^{2s} \binom{2s}{t} \sum_{a-1 \in R} (a-1)^{t-s} + 2^{2s-1} \pmod{p}.$$

Invoking Lemma 7.3, when $0 < s < \frac{p-1}{2}$, we have

$$\sum_{n \in S_+ \cap R} n^k \equiv 2^{2s-1} + \frac{1}{2} \binom{2s}{s} \frac{p-1}{2} \equiv 2^{2s-1} - \frac{1}{4} \binom{2s}{s} \pmod{p},$$

and if $\frac{p-1}{2} \le s < p - 1$, then apart from $t - s = 0, \pm \frac{p-1}{2}$, the right-hand side of the last congruence above is zero modulo p, so

$$\sum_{n \in S_+ \cap R} n^k \equiv 2^{2s-1} + \frac{p-1}{4} \left(\binom{2s}{s} + \binom{2s}{s-\frac{p-1}{2}} + \binom{2s}{s+\frac{p-1}{2}} \right)$$

$$\equiv 2^{2s-1} - \frac{1}{4} \left\{ \binom{s}{s} + 2 \binom{2s}{s-\frac{p-1}{2}} \right\} \pmod{p}.$$

This completes the proof of Theorem 7.5.

In particular, if we put $k = 1, 2$ in Theorem 7.5, we get

$$\sum_{n \in S+\cap R} n \equiv \frac{3}{2} \pmod{p},$$

$$\sum_{n \in S+\cap R} n^2 \equiv \frac{13}{2} \pmod{p}.$$

If instead we put $k = -1, -2$, then using the famous Lucas congruence, we get

$$\sum_{n \in S_+ \cap R} \frac{1}{n} \equiv \frac{1}{8} - \frac{1}{32} \left(\frac{-1}{p} \right) \pmod{p},$$

$$\sum_{n \in S_+ \cap R} \frac{1}{n^2} \equiv \frac{1}{32} - \frac{1}{2^9} \left(\frac{-1}{p} \right) \pmod{p} (p > 5).$$

In light of Theorems 7.3 and 7.5,

Therefore, we have the following corollary.

Corollary 7.2. *For $p > 3$ prime and any integer $k \equiv s \pmod{p-1}$ where $0 \le s < p - 1$,*

$$\sum_{n \in S_+} \left(\frac{n}{p} \right) n^k \equiv \begin{cases} \frac{1}{2} - \frac{1}{2} \left(\frac{-1}{p} \right) \pmod{p} & s = 0, \\ 2^{2s-1} \pmod{p} & 0 < s < \frac{p-1}{2}, \\ 2^{2s-1} - \begin{pmatrix} 2s \\ s - \frac{p-1}{2} \end{pmatrix} \pmod{p} & \frac{p-1}{2} \le s < p - 1. \end{cases}$$

In particular, for primes $p > 5$,

$$\sum_{n \in S_+} \left(\frac{n}{p} \right) n \equiv 2 \pmod{p},$$

$$\sum_{n \in S_+} \left(\frac{n}{p} \right) n^2 \equiv 8 \pmod{p},$$

$$\sum_{n \in S_+} \left(\frac{n}{p} \right) \frac{1}{n} \equiv \frac{1}{8} - \frac{1}{16} \left(\frac{-1}{p} \right) \pmod{p},$$

$$\sum_{n \in S_+} \left(\frac{n}{p} \right) \frac{1}{n^2} \equiv \frac{1}{32} - \frac{3}{256} \left(\frac{-1}{p} \right) \pmod{p}.$$

Theorem 7.6. *Fix $p > 3$ prime. Then*

$$\prod_{n \in S_+ \cap R} n \equiv \frac{3}{2} - \frac{5}{2} \left(\frac{-1}{p} \right) \pmod{p}.$$

In order to prove Theorem 7.6, we need the following lemma.

Lemma 7.4. *If p is any odd prime, then*

$$\prod_{a\in R\setminus\{1\}} (a-1) \equiv \frac{1}{2}\left(\frac{-1}{p}\right) (\mathrm{mod}\, p)$$

$$\sum_{a\in R\setminus\{1\}} \frac{1}{a-1} \equiv \frac{3}{4}(\mathrm{mod}\, p).$$

Proof. Since $1^2, 2^2, \ldots, \left(\frac{p-1}{2}\right)^2$ runs through all quadratic residues modulo p, we get from Lemma 7.2 that

$$\prod_{a\in R\setminus\{1\}} (a-1) \equiv \prod_{a=1}^{\frac{p-1}{2}} (a^2-1) \equiv \prod_{a=1}^{\frac{p-1}{2}} (a-1)(a+1)$$

$$\equiv \prod_{i=1}^{\frac{p-3}{2}} i \prod_{j=3}^{\frac{p+1}{2}} j \equiv \frac{p+1}{2(p-1)} \left(\prod_{i=1}^{\frac{p-1}{2}} i\right)^2$$

$$\equiv \frac{1}{2}(-1)^{\frac{p-1}{2}} \equiv \frac{1}{2}\left(\frac{-1}{p}\right) (\mathrm{mod}\, p)$$

and

$$\sum_{a\in R\setminus\{1\}} \frac{1}{a-1} \equiv \sum_{i=2}^{\frac{p-1}{2}} \frac{1}{i^2-1} = \frac{1}{2}\left(\sum_{i=2}^{\frac{p-1}{2}} \frac{1}{i-1} - \sum_{i=2}^{\frac{p-1}{2}} \frac{1}{i+1}\right)$$

$$= \frac{1}{2}\left(1 + \frac{1}{2} - \frac{2}{p-1} - \frac{2}{p+1}\right) \equiv \frac{3}{4}(\mathrm{mod}\, p).$$

$\square$

Proof of Theorem 7.6. From the proof of Theorem 7.4, we have

$$\prod_{n\in S_+\cap R} n \equiv \frac{1}{4} \prod_{\substack{ab\in S_+\cap R \\ ab\not\equiv 4(\mathrm{mod}\, p)}} ab \equiv \frac{1}{4} \prod_{\substack{a-1\in R \\ a-1\not\equiv\pm 1(\mathrm{mod}\, p)}} a(\mathrm{mod}\, p). \qquad (7.3)$$

If $p \equiv 1(\bmod 4)$, then both ± 1 are quadratic residues modulo p, and as $a - 1$ ranges over $R\backslash\{-1\}$, $1 - a$ ranges over $R\backslash\{1\}$. Therefore,

$$\prod_{\substack{a-1\in R \\ a-1\not\equiv\pm1(\bmod\ p)}} a \equiv \prod_{\substack{a-1\in R \\ a-1\not\equiv\pm1(\bmod\ p)}} [(a-1)+1]$$

$$\equiv \frac{1}{2} \prod_{a-1\in R\backslash\{-1\}} [(a-1)+1]$$

$$\equiv \frac{1}{2} \prod_{a-1\in R\backslash\{1\}} [-(a-1)+1]$$

$$\equiv \frac{(-1)^{\frac{p-3}{2}}}{2} \prod_{a-1\in R\backslash\{1\}} [(a-1)-1](\bmod\ p).\ (7.4)$$

Combining (7.3) and (7.3) and making use of Lemma 7.4,

$$\prod_{n\in S_+\cap R} n \equiv 4\frac{(-1)^{\frac{p-3}{2}}}{2}\frac{1}{2} \equiv -1(\bmod\ p).$$

If instead $p \equiv 3(\bmod 4)$, then 1 is a quadratic residue modulo p and -1 a quadratic nonresidue; as $a - 1$ ranges over $R, 1 - a$ ranges over N. Then

$$\prod_{\substack{a-1\in R \\ a-1\not\equiv\pm1(\bmod\ p)}} a \equiv \prod_{a-1\in R\backslash\{1\}} [(a-1)+1]$$

$$\equiv \frac{1}{2} \prod_{a-1\in R} [(a-1)+1]$$

$$\equiv \frac{1}{2} \prod_{a-1\in N} [-(a-1)+1]$$

$$\equiv \frac{(-1)^{p-1}/2}{2} \prod_{a-1\in N} [(a-1)-1](\bmod\ p).\ (7.5)$$

Invoking Lemma 7.4 and Wilson's theorem, we find that

$$\prod_{a-1\in N} [(a-1)-1] \equiv \frac{\prod_{a-1=2}^{p-1}[(a-1)-1]}{\prod_{a-1\in R\setminus\{1\}}[(a-1)-1]}$$

$$= \frac{(p-2)!}{\frac{1}{2}\left(\frac{-1}{p}\right)} \equiv -2(\bmod\, p). \qquad (7.6)$$

Then taking (7.3), (7.5), (7.6) together gives

$$\prod_{n\in S_+ \cap R} n \equiv 4\frac{(-1)^{\frac{p-1}{2}}}{2}(-2) \equiv 4(\bmod\, p).$$

This completes the proof of Theorem 7.6.

7.2   A Dual Problem

In this section, we consider the problem of solving (7.2); this is dual to the problem of solving (7.1), and we obtain various results similar to those in the previous section.

Theorem 7.7. *Let $p > 3$ be prime; then there are exactly $\frac{p-1}{2}$ distinct solutions n (modulo p) to (7.2); apart from $n \equiv 2(p-2)$ (mod p), the remaining solutions n to (7.2) always occur in pairs, given by (a,b) and $(p-b, p-a)$.*
    Let

$$S_- = \{n \mid n \equiv a - b \equiv ab(\bmod\, p)\}$$

be the set of distinct solutions modulo p to (7.2). We have the following theorem.

Theorem 7.8. *If p is an odd prime, then the product of the solutions to (7.2) is given by*

$$\prod_{n\in S_-} m \equiv -2\left(\frac{-1}{p}\right)(\bmod\, p).$$

Proof. For $2 \leq a \leq p - 1$, then equation

$$x - a - xa \equiv 0 (\bmod\, p)$$

has the unique solution

$$x \equiv -\frac{a}{a-1}(\bmod\, p).$$

In other words, $\left(a, \frac{a}{a-1}\right)\left(-\frac{a}{a-1}, a\right)$, respectively, satisfy congruences (7.1) and (7.2); therefore,

$$\prod_{n \in S} m \equiv (-1)^{\frac{p-1}{2}} \prod_{n \in S_+} n (\bmod\, p).$$

So, Theorem 7.8 follows from Theorem7.3.

Continuing to take advantage of the fact that $\left(a, \frac{a}{a-1}\right)\left(-\frac{a}{a-1}, a\right)$ are, respectively, solutions to (7.1) and (7.2), we obtain also the following theorem by comparison with Theorem 7.3. $\qquad\square$

Theorem 7.9. *If p is any odd prime, $k \equiv s(\bmod\, p - 1)$ for some $0 \leq s < p - 1$, the sum of the kth powers of all solutions to (7.2) is given by*

$$\sum_{n \in S_-} m^k \equiv \begin{cases} (-1)^s \left(2^{2s-1} - \frac{C_{2s}^s}{2}\right)(\bmod\, p) & s \neq 0, \\ \frac{p-1}{2}(\bmod\, p) & s = 0. \end{cases}$$

In particular, with $k = -1$ and -2, we have the following corollary.

Corollary 7.3. *If p is prime, then*

$$\sum_{n \in S_-} -\frac{1}{n} \equiv -\frac{1}{8} \quad (\bmod\, p) \text{ provided } p > 3,$$

$$\sum_{n \in S_-} -\frac{1}{n^2} \equiv \frac{1}{32} \quad (\bmod\, p) \text{ provided } p > 5.$$

Next, we study the intersections of S_- with R and N; we have the following result, which is similar to Theorem 7.4.

Theorem 7.10. *For p an odd prime, we have*

$$|S_- \cap R| = \frac{1}{4}\left(p - 2 + \left(\frac{-1}{p}\right)\right),$$

$$|S_- \cap N| = \frac{1}{4}\left(p - \left(\frac{-1}{p}\right)\right).$$

Proof. Consider any n contained in S_-, in other words satisfying $n \equiv a - b \equiv ab \pmod{p}$; then $(a + 1)(1 - b) \equiv 1 \pmod{p}$. Supposing $a + 1 \equiv 1 - b \pmod{p}$, if $a + 1 \equiv 1 - b \equiv 1 \pmod{p}$, then $a \equiv b \equiv 0 \pmod{p}$, $n \equiv ab \equiv 0 \pmod{p}$, contradicting the hypothesis that $n \in S_-$. So in this case $n \in S_-$ only if $a + 1 \equiv 1 - b \equiv -1 \pmod{p}$, in which case $a \equiv -2, b \equiv 2 \pmod{p}$, $n \equiv ab \equiv -4 \pmod{p}$.

Now if in fact $n \in S_- \cap R$, then $\left(\frac{n}{p}\right) = 1$. Noting that $n \equiv ab \equiv \frac{b^2}{1-b} \pmod{p}$, we see that also both $\left(\frac{a-1}{p}\right) = 1, \left(\frac{1-b}{p}\right) = 1$, and similarly both $\left(\frac{b-1}{p}\right) = 1, \left(\frac{a+1}{p}\right) = 1$. So $|S_- \cap R|$ consists of distinct pairs $(a + 1, 1 - b)$ modulo p satisfying $(a+1)(1-b) \equiv 1 \pmod{p}$ and $\left(\frac{a+1}{p}\right) = \left(\frac{1-b}{p}\right) = 1$. Apart from the case $a+1 \equiv 1-b \equiv \pm 1 \pmod{p}$, the two components of every other pair are distinct modulo p.

If $p \equiv 1 \pmod 4$, then both ± 1 are quadratic residues modulo p, so

$$|S_- \cap R| = \frac{\frac{p-1}{2} - 2}{2} + 1 = \frac{p-1}{4} = \frac{1}{4}\left(p - 2 + \left(\frac{-1}{p}\right)\right);$$

if $p \equiv 3 \pmod 4$, then 1 is a quadratic residue and -1 a quadratic non-residue, so

$$|S_- \cap R| = \frac{\frac{p-1}{2} - 1}{2} = \frac{p-3}{4} = \frac{1}{4}\left(p - 2 + \left(\frac{-1}{p}\right)\right).$$

Finally,

$$|S_- \cap N| = |S_-| - |S_- \cap R| = \frac{1}{4}\left(p - \left(\frac{-1}{p}\right)\right).$$

This completes the proof of Theorem 7.10. $\qquad\square$

Theorem 7.11. *If $p > 3$ is prime, $k \equiv s \pmod{p-1}$ for some $0 \le s < p - 1$, then*

$$\sum_{n \in S - \cap R} n^k$$

$$\equiv \begin{cases} -\frac{1}{2} + \frac{1}{4}\left(\frac{-1}{p}\right) \pmod{p} & s = 0, \\[2mm] (-1)^s\left(1 + \left(\frac{-1}{p}\right)\right)2^{2s-2} \qquad -\frac{(-1)^s}{4}\binom{2s}{s}\pmod{p}, \\[2mm] (-1)^s\left(1 + \left(\frac{-1}{p}\right)\right)2^{2s-2} - \frac{(-1)^s}{4} & 0 < s < \frac{p-1}{2}, \\[2mm] \left\{\binom{2s}{s} + 2\left(\frac{-1}{p}\right)\binom{2s}{s-\frac{p-1}{2}}\right\}\pmod{p} & \frac{p-1}{2} \le s < p - 1. \end{cases}$$

Proof. When $s = 0$, the claim follows directly from Theorem 7.4. Supposing otherwise that $0 < s < p - 1$, for $1 < b \le p - 1$, we have

$$a \equiv \frac{b}{1-b} \equiv \frac{a}{a-1}\pmod{p}.$$

Except in the case that $n = 2(p-2)$, the other (a, b) solving (7.2) must all satisfy $a \not\equiv -b \pmod{p}$. When $p \equiv 1 \pmod 4$, $n = 2(p-2)$ does not lie in $S_- \cap R$, and when $p \equiv 3 \pmod 4$, $n = 2(p-2)$ does lie in $S_- \cap R$. Therefore when $p \equiv 1 \pmod 4$,

$$\sum_{n \in S - \cap R} n^k \equiv \sum_{n \in S_- \cap R} n^s \equiv \frac{1}{2}\left(\sum_{1-b \in R} \frac{b^{2x}}{(1-b)^s} - (-2)^{2s}\right) + (-2)^{2s}$$

$$\equiv \frac{1}{2}\sum_{1-b \in R} \frac{b^{2s}}{(1-b)^s} + (-1)^s 2^{2s-1}\pmod{p},$$

and when $p \equiv 3 \pmod 4$,

$$\sum_{n \in S - \cap R} n^s \equiv \frac{1}{2}\sum_{1-b \in R} \frac{b^{2s}}{(1-b)^s}\pmod{p}.$$

So for $p > 3$, $0 < s < p - 1$,

$$\sum_{n \in S_- \cap R} n^s \equiv \frac{1}{2}\left(\sum_{1-b \in R} \frac{b^{2s}}{(1-b)^s}\right) + (-1)^s\left(1 + \left(\frac{-1}{p}\right)\right)2^{2s-2}$$

$$\equiv \frac{1}{2}\left(\sum_{1-b \in R} \frac{(1-b-1)^{2s}}{(1-b)^s}\right) + (-1)^s\left(1 + \left(\frac{-1}{p}\right)\right)2^{2s-2}$$

$$\equiv \frac{1}{2}\left(\sum_{1-b \in R}\sum_{t=0}^{2s}(-1)^t\binom{2s}{t}(1-b)^{t-s}\right)$$

$$+ (-1)^s\left(1 + \left(\frac{-1}{p}\right)\right)2^{2s-2}$$

$$\equiv \frac{1}{2}\left(\sum_{t=0}^{2s}(-1)^t\binom{2s}{t}\sum_{1-b \in R}\sum_{t=0}^{2s}(-1)^t\binom{2s}{t}\right)$$

$$+ (-1)^s\left(1 + \left(\frac{-1}{p}\right)\right)2^{2s-2} \pmod p.$$

By Lemma 7.3, if $0 < s < \frac{p-1}{2}$, then

$$\sum_{n \in S_- \cap R} n^k \equiv (-1)^s\left(1 + \left(\frac{-1}{p}\right)\right)2^{2s-2} + \frac{(-1)^s}{4}\binom{2s}{s}\frac{p-1}{2}$$

$$\equiv (-1)^s\left(1 + \left(\frac{-1}{p}\right)\right)2^{2s-2} - \frac{(-1)^s}{4}\binom{2s}{s} \pmod p$$

and if $\frac{p-1}{2} \le s < p-1$, then apart from $t - s = 0, \pm\frac{p-1}{2}$, the remaining terms in the first sum of all zero modulo p, so

$$\sum_{n \in S_- \cap R} n^k \equiv (-1)^s\left(1 + \left(\frac{-1}{p}\right)\right)2^{2s-2} + \frac{p-1}{4}(-1)^s\binom{2s}{s}$$

$$+ 2(-1)^{s-\frac{p-1}{2}}\binom{2s}{s - \frac{p-1}{2}}$$

$$\equiv (-1)^s \left(1 + \left(\frac{-1}{p}\right)\right) 2^{2s-2} - \frac{(-1)^s}{4} \left(\binom{2s}{s}\right)$$

$$+ 2 \left(\frac{-1}{p}\right) \left(\binom{2s}{s - \frac{p-1}{2}}\right) \pmod{p}.$$

This completes the proof of Theorem 7.11. □

In particular, with $k = -1$ and -2, we get

$$\sum_{n \in S_- \cap R} \frac{1}{n^2} \equiv \frac{1}{32} - \frac{1}{2^9} \left(\frac{-1}{p}\right) \pmod{p}(p > 5),$$

$$\sum_{n \in S_- \cap R} \frac{1}{n^2} \equiv \frac{1}{64} \left(\frac{-1}{p}\right) + \frac{5}{2^9} \pmod{p}(p > 5).$$

Using Theorems 7.9, 7.11, and the identity

$$\sum_{n \in S_-} \left(\frac{n}{p}\right) n^k = \sum_{n \in S_- \cap R} n^k - \sum_{n \in S_- \cap N} n^k = 2 \sum_{n \in S_- \cap R} n^k - \sum_{n \in S_-} n^k$$

we get the following corollary.

Corollary 7.4. *For $p > 3$ prime and an integer $k \equiv s \pmod{p-1}$ for some $0 \le s < p - 1$,*

$$\sum_{n \in S_-} \left(\frac{n}{p}\right) n^k$$

$$\equiv \begin{cases} \frac{1}{2} \left(\frac{-1}{p}\right) - \frac{1}{2} \pmod{p} & \text{if } s = 0, \\[2mm] (-1)^k \left(\frac{-1}{p}\right) 2^{2s-1} \pmod{p} & \text{if } 0 < s < \frac{p-1}{2}, \\[2mm] (-1)^k \left(\frac{-1}{p}\right) \left(2^{2s-1} - \left(\binom{2s}{s - \frac{p-1}{2}}\right)\right) \pmod{p} \\[2mm] \hspace{3cm} \text{if } \frac{p-1}{2} \le s < p - 1. \end{cases}$$

In particular, for $p > 5$, setting $k = \pm 1, \pm 2$, we get

$$\sum_{n \in S_-} \left(\frac{n}{p}\right) n \equiv 2 \left(\frac{-1}{p}\right) \pmod{p},$$

$$\sum_{n \in S_-} \left(\frac{n}{p}\right) n^2 \equiv 8 \left(\frac{-1}{p}\right) \pmod{p},$$

$$\sum_{n \in S_-} \left(\frac{n}{p}\right) \frac{1}{n} \equiv -\frac{1}{8} \left(\frac{-1}{p}\right) \pmod{p},$$

$$\sum_{n \in S_-} \left(\frac{n}{p}\right) \frac{1}{n^2} \equiv \frac{1}{32} \left(\frac{-1}{p}\right) \pmod{p}.$$

Theorem 7.12. *For $p > 3$ prime,*

$$\prod_{n \in S_- \cap R} n \equiv -\frac{1}{4} \left(\frac{2}{p}\right) - \frac{3}{4} \left(\frac{-2}{p}\right) \pmod{p}.$$

Proof. If $p \equiv 1 \pmod 4$, then ± 1 are both quadratic residues modulo p; from the proofs of Theorems 7.10 and Lemma 7.4, we have

$$\prod_{n \in S_- \cap R} n \equiv -4 \prod_{\substack{ab \in S_- \cap R \\ ab \not\equiv 4 \pmod p}} ab$$

$$\equiv (-1)^{\frac{p-1}{4}} 4 \prod_{1-b \in R \setminus \{1, -1\}} [(1-b) - 1]$$

$$\equiv (-1)^{\frac{p-5}{4}} 2 \prod_{1-b \in R \setminus \{1\}} [(1-b) - 1]$$

$$\equiv (-1)^{\frac{p-5}{4}} 2 \cdot \frac{1}{2} \left(\frac{-1}{p}\right) = (-1)^{\frac{p+3}{4}} \pmod{p}.$$

If $p \equiv 3 \pmod 4$, then 1 is a quadratic residue modulo p, -1 a quadratic non-residue, and we have

$$\prod_{n \in S_- \cap R} n \equiv \prod_{ab \in S_- \cap R} ab$$

$$\equiv (-1)^{\frac{p-1}{4} - \frac{1}{2}} \prod_{1-b \in R \setminus \{1\}} [(1-b) - 1]$$

$$\equiv (-1)^{\frac{p-3}{4}} \frac{1}{2} \left(\frac{-1}{p}\right) = \frac{(-1)^{\frac{p+1}{4}}}{2} \pmod{p}.$$

This completes the proof of Theorem 7.12. $\qquad \square$

Figure 7.1. Jacques Hadamard.

We next relate the various topics discussed above to a special class of matrices (see Cai, 2021), known as Hadamard (Fig. 7.1) matrices after Jacques-Salomon Hadamard (1865–1963), a French mathematician who independently proved the prime number theorem in 1896. Incidentally, Hadamard made two trips to China in 1936, visiting Shanghai, Beijing, and Hangzhou. The Hadamard matrix problem is a very famous topic in combinatorics, with applications to errorcorrecting codes in communication systems, for example to help make broadcast images of landing on the moon clearer. Later, we connect the problem of constructing Hadamard matrices with the discussion of the solvability of the above additive and multiplicative congruences.

A real square matrix $A = \{a_{ij}\}_{n \times n}$ of order n is a Hadamard matrix if every $a_{ij} = \pm 1$, and the rows (or columns) are mutually orthogonal. It is easy to see that in this case $A^{A^T} = nI_n$ and $|A| = \pm(\sqrt{n})^n$. Moreover, the order of a Hadamard matrix can only be 1, 2 or a multiple of 4. It is straightforward enough to construct Hadamard matrices of orders 1 and 2, but Hadamard matrices of other orders are not so easy to find.

Hadamard Conjecture. There exists a Hadamard matrix of order $4n$ for every positive integer n.

In 1933, the British mathematician Raymond Paley (1907–1933) discovered an effective construction for Hadamard matrices making use of the theory of quadratic residues.

Proposition. *Let p be a prime number of the form $4k+3$, R and N respectively the sets of quadratic residues and quadratic nonresidues modulo p; then define a square matrix B of order p as follows: the entry $b_{ij} = 1$ if $j - i \in R$ and -1 if $j - i \in N$. Set A to be the square matrix of order $p + 1$ with every entry in the first row and first column equal to 1 and the remaining order p square submatrix given by B. Then A is a Hadamard matrix of order $p + 1$.*

It is not difficult to prove this proposition, using the theory of quadratic residues and the Legendre symbol. By a similar method, Paley also provided a construction for Hadamard matrices of order $2(q + 1)$ where q is a prime of the form $4k + 1$. It can be said that Paley has made the greatest contribution to date to the topic of Hadamard matrix existence results. The smallest multiple of 4 for which it has not yet been shown that a Hadamard matrix exists is 668; there are in total twelve more multiples of 4 not exceeding 2000 for which no Hadamard matrix has yet been discovered; these are $716, 892, 1004, 1132, 1244, 1388, 1436, 1676, 1772, 1916,$ 1948, 1964.

Since the number of elements in the solution sets S_+ for (7.1) and S_- for (7.2) is given by $\frac{p-1}{2}$, which as the same as the size of the sets of quadratic residues R and quadratic nonresidues N, we are inclined to explore the following new possibility: whether or not the definitions and properties of S_+ and S_- can be used to construct a new Hadamard matrix?

7.3   Catalan's Conjecture

In 1343, the French mathematician, astronomer, philosopher, and theologian Gersonides (1288–1344) wrote *The Harmony of Numbers*, in which he arrived at the following conclusion:

if x, y are drawn from the numbers 2, 3, and $a, b > 1$ are integers, then the only solution to the equation

$$x^a - y^b = 1 \tag{7.3}$$

is given by $(x, y; a, b) = (3, 2; 2, 3)$.

Figure 7.2. Portrait of Catalan.

Gersonides, whose father was a Jewish writer named Gerson ben Solomon Catalan (Fig. 7.2), was one of the earliest French mathematicians to leave his name to history; Gersonides is the Graecized version of his name, which was actually Levi ben Gershom. Prior to witing *The Harmony of Numbers*, he also published a mathematical treatise entitled *Maaseh Hoshev* (1321) and another *On Sines, Chords, and Arcs* (1342). The first of these discusses arithmetic, including the extraction of square and cube roots, while the latter deals with trigonometry, in particular the sine law for plane triangles, and included sine tables of five figures. In addition to his mathematical works, Gersonides also published a criticism of some arguments due to Aristotle in 1319, showing the influence of the Andalusian (Spanish Islamic) writer Averroes (1126–1198), one of the two most important Islamic philosophers alongside Avicenna (980–1037).

This conclusion languished in obscurity for some five centuries until 1844, when the French–Belgian mathematician Eugene Charles Catalan (1814–1894) revived and generalized it, allowing x, y to take any values in positive integers. In a letter to the editor of *Crelle's Journal* in Berlin, he proposed the following conjecture.

Catalan's Conjecture. There are no consecutive positive integers of the form $\{x^m, y^n\}$ where $m, n > 1$ are positive integers other 8,9.

Catalan was 30 years old when he made this conjecture. He was born in Bruges in Belgium (at that time a part of France), and his family had moved to Paris around 1825. He was the only son of a jeweler, but became interested in mathematics and graduated from the famous École Polytechnique in Paris, where the mathematician Joseph Liouville (1809–1892) was his teacher. He subsequently stayed on at the same school as an assistant professor, serving as a mentor to Charles Hermite (1822–1901), who later proved that e is a transcendental number.

In addition to this conjecture, Catalan made numerous other mathematical discoveries; for example, there are the Catalan numbers, defined by

$$C_n = \frac{1}{n+1}\binom{2n}{n}(n \geq 0).$$

The first 10 Catalan numbers are $1, 1, 2, 5, 14, 42, 132, 429, 1430,$ $4862, \ldots$, and the numbers in this sequence occur often in counting problems, playing an important role in combinatorics. It is worth a mention that these numbers in fact appear also in a book entitled *The Quick Method for Obtaining the Precise Ratio of Division of a Circle*, by Minggatu (or Ming Antu, 1692–1763), a Mongolian mathematician active in China during the Qing dynasty. For this reason, some scholars have suggested that this sequence should be known instead as the Minggatu-Catalan numbers.

In 1865, Catalan was hired as a professor at the University of Liège in Belgium, where he stayed on until his retirement at the age of seventy. In 1879, when he was 65 years old, Catalan discovered an identity between terms of the famous Fibonacci sequence:

$$F_n^2 - F_{n-r}F_{n+r} = (-1)^{n-r}F_r^2 (n > r \geq 1)$$

now known as the Catalan identity. In the case that $r = 1$, this becomes

$$F_{n-1}F_{n+1} - F_n^2 = (-1)^n (n \geq 1),$$

which is the Cassini identity, discovered two centuries earlier by Giovanni Cassini (1625–1712), then the director of the Paris Observatory. There are various ways to prove it, including by induction and by the use of matrices.

Returning to Catalan's conjecture, Euler proved that it holds in the case that $m = 3, n = 2$. In 1850, the French mathematician Victor-Amédée Lebesgue (1791–1875, not to be confused with the later mathematician Henri Lebesgue, the founder of modern real analysis, who lived from 1875 to 1941) resolved the $m = 2$ case, therefore also the case of any even m. In 1932, Atle Selberg resolved the $n = 4$ case, and in 1962 Ke Zhao (1910–2002) resolved the $n = 2$ case, therefore also the case of any even n. The best result for odd values of m, n however, was always the case that $m = n = 3$, settled by the Norwegian mathematician Trygve Nagell (1895–1988) in 1921. In 1976, the Dutch mathematician Robert Tijdeman (1943–) proved (see Tijdeman, 1976) that the Catalan equation has at most finitely many solutions.

As in the case of Fermat's last theorem (Fig. 7.3), for which it is sufficient to consider only odd primes p in place of the exponent $n \geq 3$ in (2.2), it is sufficient to prove Catalan's conjecture to consider only

Figure 7.3. Issac Newton Mathematics Institute, University of Cambridge, the place where Fermat's Last Theorem was declared to be proved. Photograph by the author.

prime numbers p, q in place of the exponents a, b in (7.3); in other words, to show that the equation

$$x^p - y^q = 1 \qquad (7.4)$$

has no solutions in positive integers for any primes p, q. We show below how to prove that (7.4) has no solutions when $q = 2$, and only one solution when $p = 2$ (see also Ke, 1962 and Ke and Sun, 2011).

Theorem 7.13. *Let p be an odd prime; then the equation*

$$y^2 + 1 = x^p \qquad (7.5)$$

has no solutions in positive integers.

Proof. It is easy to see that in any solution to (7.5) in positive integers, one of x, y must be odd and the other even. If x is even and y odd, then from reducing (7.5) implies that $2 \equiv 0 \pmod 8$, absurd. So we must have x odd, y even. We factor the left-hand side of (7.5) in the ring of gaussian integers to get

$$(1 + yi)(1 - yi) = x^p. \qquad (7.6)$$

Suppose $(1 + yi, 1 - yi) = \alpha$; then for any gaussian prime $\beta \mid \alpha$, we must have $\beta \mid 2$, and therefore $\beta \mid y$. Then since $\beta \mid 1 + yi$, it follows that $\beta \mid 1$, which is possible, since β is not a unit. We conclude that $(1 + yi, 1 - yi) = 1$. Since the ring of gaussian integers is a unique factorization domain,

$$1 + yi = i^r(u + iv)^p, \quad 0 \le r \le 3, \quad x = u^2 + v^2. \qquad (7.7)$$

It is easy to see that for $0 \le r \le 3$ (considering separately the cases where p is congruent to 1 or 3 modulo 4 when $r = 1$ or 3), the unit i^r in (7.7) can be incorporated into the following term $(u + iv)^p$, so that we have simply

$$1 + yi = (u + iv)^p, \quad x = u^2 + v^2. \qquad (7.8)$$

This expands as

$$1 + yi = u^p + pu^{p-1}iv - \frac{p(p-1)}{2}u^{p-2}v^2 + \cdots \pm pu v^{p-1} \pm iv^p$$

and comparing the real parts on either side, we see that

$$1 = u^p - \frac{p(p-1)}{2}u^{p-2}v^2 + \cdots \pm puv^{p-1}. \qquad (7.9)$$

Therefore, $u|1$, that is, $u = \pm 1$, and since x is odd, so that one of u, v is odd and the other even, it must be the case that v is even. In light of this and (7.9), we find that $u = 1$, and substituting into (7.9) gives

$$\frac{p(p-1)}{2} - \frac{p(p-1)(p-2)(p-3)}{4!} v^2$$
$$+ \cdots + p v^{p-3} = 0. \tag{7.10}$$

For $k \geq 1$, the terms of (7.10) can be rewritten as

$$(-1)^{k-1} \binom{p}{2k} v^{2k-2} = (-1)^{k-1} \binom{p}{2} \binom{p-2}{2k-2} \frac{2v^{vk-2}}{(2k-1)2k}. \tag{7.11}$$

Then since v is an even number, we find that for $k > 1$ the power of two in the numerator of the last part of each term is greater than the power of 2 in the $k = 1$ term, so that (7.10) cannot hold. This contradiction completes the proof of Theorem 7.13. $\qquad\square$

Theorem 7.14. *If p is an odd prime, then the equation*

$$y^2 - 1 = x^p \tag{7.12}$$

has a solution only if $p = 3$, in which case the only solution is $(x, y) = (2, 3)$.

In order to prove Theorem 7.14, we need the following lemma (see Ke and Sun, 2011).

Lemma 7.5. *There exist positive integer solutions to (7.4) if and only if*

$$x + 1 = p^{sq-1} y_1^q, \quad \frac{x^p + 1}{x + 1} = p y_2^q, \quad y = p^s y_1 y_2,$$

$$(y_1, y_2) = 1, p \nmid y_1 y_2, \tag{7.13}$$

$$y - 1 = q^{tp-1} x_1^p, \quad \frac{y^q - 1}{y - 1} = q x_2^p, \quad x = q^t x_1 x_2,$$

$$(x_1, x_2) = 1, \quad q \nmid x_1 x_2, \tag{7.14}$$

where s, t, x_1, x_2, y_1, y_2 are all positive integers.

Proof of Theorem 7.14. From (7.13) in Lemma 7.6, we get

$$x + 1 = pp^{2s-1}y_1^2, \quad \frac{x^p + 1}{x + 1} = py_2^2, \quad y = p^s y_1 y_2. \tag{7.15}$$

From the second identity in (7.14), with $q = 2$, we see that any y satisfying (7.12) must be odd, so that x must be even.

We suppose first that $p \equiv 5$ or $7 \pmod 8$, specifically $p = 8u + a$ with $a = 5$ or 7. By the first equation in (7.15), $x \equiv a - 1 \pmod 8$, so that (7.12) can be rewritten as

$$y^2 = (x^2 - 1 + 1)^{4u} x^a + 1 \equiv x^a + 1 \pmod{x^2 - 1}.$$

By the quadratic reciprocity law for the Jacobi symbol,

$$\left(\frac{x^x + 1}{x - 1}\right) = \left(\frac{2}{x - 1}\right) = 1. \tag{7.16}$$

But this contradicts the fact that $x - 1 \equiv 3$ or $5 \pmod 8$.

Then suppose $p \equiv 3 \pmod 8$. If actually $p = 3$, then Gersonides had already discovered that the only set of positive integer solutions is $(x, y) = (2, 3)$. For $p > 3$, we can write $p = 8u + 3 = 24v + a$ with $a = 11$ or 19. We have

$$y^2 = (x^3 - 1 + 1)^{8v} x^a + 1 \equiv x^a + 1 \pmod{x^3 - 1}.$$

Therefore,

$$\left(\frac{x^a + 1}{x^3 - 1}\right) = 1 \tag{7.17}$$

If $a = 11, x \equiv 2 \pmod 8$, noting that

$$x^{11} - x^2 = x^2 \left(x^9 - 1\right)$$

and making use of the properties of the Jacobi symbol, we get

$$\left(\frac{x^a + 1}{x^3 - 1}\right) = \left(\frac{x^2 + 1}{x^3 - 1}\right) = \left(\frac{x^3 - 1}{x^2 + 1}\right) = \left(\frac{-x - 1}{x^2 + 1}\right) = \left(\frac{x^2 + 1}{x + 1}\right)$$

$$= \left(\frac{2}{x + 1}\right) = -1$$

contradicting (7.17).

If $a = 19, x \equiv 2 \pmod 8$, then

$$1 = \left(\frac{x^{19}+1}{x^3-1}\right) = \left(\frac{x+1}{x^3-1}\right) = -\left(\frac{x^3-1}{x+1}\right) = \left(\frac{2}{x+1}\right) = -1$$

again a contradiction.

The only remaining case is $p \equiv 1 \pmod 8$, in which case $x \equiv 0 \pmod 8$; according to the second identity in (7.15),

$$p^{y_2^2} = x^{p-1} - y^{p-2} + \cdots - x + 1, \tag{7.18}$$

Suppose $1 < l < p$ with l odd, $p = kl + a$ where $0 < a < l$, $(a, l) = 1$,

$$E(t) = \frac{(-x)^2 - 1}{(-x)^2 - 1}, \quad t \geq 1 \geq 1,$$

Since $x \equiv 0 \pmod 8$, therefore $E(t) \equiv 1 \pmod 8$. From the second identity in (7.15),

$$p y_2^2 = \frac{x^p + 1}{x+1} = \frac{x^{kl+a}+1}{x+1}. \tag{7.19}$$

Since $x^l + 1 = (x+1)E(l)$, $k + l \equiv 1 \pmod 2$ (7.19) becomes

$$p y_2^2 = \frac{((x+1)E(l) - 1)^k x^a + 1}{x+1} \equiv \frac{(-1)^k x^a + 1}{x+1} = E(a) \pmod{E(l)}. \tag{7.20}$$

Note that

$$(E(a), E(1)) = \frac{(-x)^{(a-1)}-1}{(-x)-1} = 1.$$

We prove that

$$\left(\frac{E(a)}{E(l)}\right) = 1. \tag{7.21}$$

By the Euclidean algorithm,

$$1 = k_1 a + r_1, 0 < r_1 < a,$$

$$a = k_2 r_1 + r_2, 0 < r_2 < r_1,$$

$$r_1 = k_3 r_2 + r_3, 0 < r_3 < r_2,$$

$$\cdots$$

$$r_{s-1} = k_{s+1} r_s + r_{s+1}, 0 < r_{s+1} < r_s,$$

$$r_s = k_{s+2} r_{s+1}.$$

Since

$$\frac{(-x)^{k_1x+r_1}-1}{-x-1} \equiv \frac{(-x)^{r_1}-1}{-x-1}\left(\bmod \frac{(-x)^{x^n}-1}{-x-1}\right),$$

or

$$E\left(k_1a+r_1\right) \equiv E\left(r_1\right)(\bmod E(a)),$$

from $(1,p)=1$, we see that $r_{s+1}=1$. Therefore

$$\left(\frac{E(a)}{E(l)}\right) = \left(\frac{E(l)}{E(a)}\right) = \left(\frac{E\left(k_1a+r_1\right)}{E(a)}\right) = \left(\frac{E\left(r_1\right)}{E(a)}\right)$$

$$= \left(\frac{E(a)}{E\left(r_1\right)}\right) = \left(\frac{E\left(r_2\right)}{E\left(r_1\right)}\right) = \cdots = \left(\frac{E\left(r_{s-1}\right)}{E\left(r_s\right)}\right)$$

$$= \left(\frac{E\left(r_{s+1}\right)}{E\left(r_1\right)}\right) = \left(\frac{E(1)}{E\left(r_s\right)}\right) = \left(\frac{1}{E\left(r_s\right)}\right) = 1,$$

which proves (7.21).

By (7.20) again

$$1 = \left(\frac{pE(a)}{E(l)}\right) = \left(\frac{p}{E(l)}\right). \tag{7.22}$$

According to the second identity in (7.15), $x \equiv -1(\bmod p)$, so

$$\left(\frac{p}{E(l)}\right) = \left(\frac{E(l)}{p}\right) = \left(\frac{l}{p}\right).$$

Then since $p \equiv 1(\bmod 8), p \geq 17$, we can take l a quadratic non-residue modulo p, in other words so that

$$\left(\frac{l}{p}\right) = -1$$

contradicting (7.22). This completes the proof of Theorem 7.14.

Corollary 7.5. *The diophantine equation*

$$x(x+1)(x+2)\cdots(x+n-1) = y^k, k > 1 \tag{7.23}$$

has no solutions in non-zero integers when $n = 3$ or 4.

When $n = 3$, we have

$$x(x+1)(x+2) = y^k, \quad k > 1. \tag{7.24}$$

Since $(x(x+2), x+1) = 1$, therefore

$$x + 1 = u^k, x(x+2) = v^k, \quad y = uv. \tag{7.25}$$

It is easy to see that the second identity in (7.25) is equivalent to

$$(x+1)^2 - 1 = v^k.$$

Then by Theorem 7.14, the only solutions to this equation are $v = 2, k = 3, x = 2$ or -4. None of these satisfy the first identity in (7.25), namely $x + 1 = u^3$, so (7.24) has no solutions.

When $n = 4$, we have

$$x(x+1)(x+2)(x+3) = y^k, \quad k > 1, \tag{7.26}$$

equivalently

$$\left(x^2 + 3x + 1\right)^2 - 1 = y^k, \quad k > 1. \tag{7.27}$$

According to Theorem 7.14, the only solution to (7.27) is $y = 2$, $k = 3, x^2 + 3x + 1 = \pm 3$. This last equation is obviously impossible, so that (7.26) has no non-zero integer solutions. This completes the proof of Corollary 7.5.

Paul Erdős and John Selfridge proved (see Erdős-Selfridge, 1975) in 1975 that (7.23) has no nonzero solutions when $n > 1$.

In 2002, the Romanian-German mathematician Preda Mihăilescu (1955–) finally proved (see Mihăilescu, 2004) Catalan's conjecture, using the theory of cyclotomic fields and Galois modules, so that now this conjecture is also known as Mihăilescu's theorem. Intriguingly, his proof also made use of Wieferich pairs. Prior to discussing Wieferich pairs, we should first say a bit about Wieferich primes. In 1909, the German mathematician Arthur Wieferich (1884–1954) asked whether there exists any prime number p satisfying

$$2^{p-1} \equiv 1 \left(\bmod p^2\right).$$

Such primes are referred to as Wieferich primes. So far, searching among all numbers not exceeding 6.7×10^{15}, there are only two known Wieferich primes, which are 1093 and 3511, respectively discovered in 1913 and 1922. Nevertheless, nobody has been able to prove that there exist infinitely many primes that are not Wieferich primes.

In 1988, the American mathematician Joseph H. Silverman (1955–) proved the latter claim holds under the *abc* conjecture that is, the for infinitely many primes p, we have $2^{p-1} \not\equiv 1 \pmod{p^2}$. On the other hand, whether or not there are infinitely many Wieferich primes remains a complete mystery.

Wieferich also considered pairs p, q of prime numbers satisfying

$$p^{q-1} \equiv 1 \pmod{q^2}, \quad q^{p-1} \equiv 1 \pmod{p^2},$$

which are known as Wieferich pairs. To date there have been discovered seven Wieferich pairs: (2, 1093), (3, 1006003), (5, 1645333507), (5, 188748146801), (83, 4871), (911, 318917), (2903, 18787). In his proof of Catalan's conjecture, Mihăilescu obtained several necessary conditions, among them that the exponents p and q in (7.4) form a Wieferich pair.

We consider next a strong version of the Catalan equation: $a - b = 1$ where a and b are square-full numbers. Here, a positive integer n is called square-full if p^2 divides n whenever a prime number p divides n. It is easy to see that this is equivalent to $a - b = 1$ with ab a squarefull number. Using the known properties of Pell's equation, it is not difficult to show that there are infinitely many solutions, for example $(8, 9), (288, 289), (675, 676)$, and so on.

On the other hand, if we call a positive integer n cube-full if p^3 divides n whenever a prime p divides n, then after numerical calculations we conjecture that, for any positive integer s, the only pair (a, b) of positive integers satisfying $a - b = 2^s$ with ab cube-full is $(2^{s+1}, 2^s)$. □

7.4 New Egyptian Fractions

Finally, bringing both this chapter and this book to a close, we consider one last example of additive and multiplicative equations.

The ancient Egyptians liked to work with unit fractions, that is, rational numbers with numerator 1 (Fig. 7.4). The Rhind Papyrus, one of the most ancient extant mathematical documents, discusses

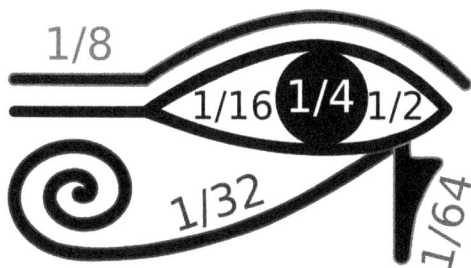

Figure 7.4. Egyptian fractions: Eyes of the Gods.

how to express a positive rational number as a sum of unit fractions, in other words to find an expression of the form

$$\frac{m}{n} = \frac{1}{x_1} + \frac{1}{x_2} + \cdots + \frac{1}{x_k}. \tag{7.28}$$

for given positive integers m, n.

It is always possible to obtain such an expression, but various problems arise under certain special restrictions. If $1 \leq m \leq 3$, $k = 3$, then it is easy to see that (7.28) has solutions for any positive integer n. If $m = 4, k = 3$, then we have the 1948 conjecture (see Guy, 2004) due to Paul Erdös (1913–1996) and the German-born American mathematician Ernst G. Straus (1922–1983), at one point assistant to Albert Einstein, that the equation

$$\frac{4}{n} = \frac{1}{x} + \frac{1}{y} + \frac{1}{z} \tag{7.29}$$

has a solution for every $n > 1$. This conjecture is valid when $n \equiv 2 \,(\mathrm{mod}\ 3)$, since in this case

$$\frac{4}{n} = \frac{1}{n} + \frac{1}{\frac{n-2}{3} + 1} + \frac{1}{n\left(\frac{n-2}{3} + 1\right)}.$$

In 1969, Mordell proved (see Mordell, 1969) that it is also valid when $n \neq 1 (\mathrm{mod}\ 24)$. Numerically, the conjecture has so far been verified for all $n \leq 10^{14}$.

In 1970, William A. Webb proved (see Webb, 1970, p. 70) that (7.29) has a solution for almost all integers $n > 1$; in fact, he proved the even stronger result that

$$S(N) \ll \frac{N}{\log^{7/4} N},$$

where $S(N)$ indicates the number of n not exceeding N for which the conjecture fails. In 1982, Yang Xun Qian proved (see Yang, 1987) the following theorem.

Theorem 7.15. *With $S(N)$ as above,*

$$S(N) \ll \frac{N}{\log^2 N}.$$

In order to prove Theorem 7.5, we introduce the following lemma (see Halberstam and Richert, 1974).

Lemma 7.6. *Let g be a positive integer, a_i and b_i coprime positive integers for $1 \le i \le g$, and define*

$$E = \sum_{i=1}^{g} a_i \prod_{1 \le r < s \le g} (a_r b_s - a_s b_r) \neq 0.$$

Suppose also that y and x are real numbers with $1 \le y \le x$ and $\mathcal{B}$ is a family of prime numbers, such that

$$\sum_{\substack{p < y \\ p \in \mathcal{B}}} \frac{1}{p} \ge \delta \log \log y - A$$

for some positive constants δ and A. Then

$$|\{n : x - y < n \le x, ((a_i n + b_i), \mathcal{B}) = 1, \ 1 \le i \le s\}|$$

$$\ll \prod_{p | E, p \in \mathcal{B}} \left(1 - \frac{1}{p}\right)^{\rho(p) - g} \frac{y}{\log^{\delta_g} y}. \tag{7.30}$$

Here, $\delta(p)$ denotes the number of solutions to the equation

$$\sum_{i=1}^{g} (a_i n + b_i) \equiv 0 \,(\mathrm{mod}\, p)$$

and the constant implied by the symbol $\ll$ depends only on g and A.

Proof of Theorem 7.15. Obviously,

$$\frac{4}{n} = \begin{cases} \frac{1}{nk(k+1)} + \frac{1}{n(k+1)} + \frac{1}{kv}, & n = (4k-1)v, \\ \frac{1}{nk} + \frac{1}{nkv} + \frac{1}{kv}, & n+1 = (4k-1)v, \\ \frac{1}{nk} + \frac{1}{nk(kv-1)} + \frac{1}{kv-1}, & n+4 = (4k-1)v, \\ \frac{1}{nk} + \frac{1}{k(kv-n)} + \frac{1}{n(kv-n)}, & 4n+1 = (4k-1)v. \end{cases}$$

It follows that the conclusion holds whenever one of $n, n+1, n+4$ or $4n+1$ has a factor $d = 4k-1$.

Now take

$$\mathcal{B} = \{p : p \equiv -1 (\mathrm{mod}\ 4)\}, \quad y = x, \quad g = 4,$$

$$\prod_{i=1}^{4} (a_i x + b_i) = x(x+1)(x+4)(4x+1)$$

in Lemma 7.6. We get

$$E = 2^4 \cdot 3^3 \cdot 5 \neq 0, \quad \delta(3) = 2,$$

$$\prod_{p \mid E \in p \in \mathcal{B}} \left(1 - \frac{1}{p}\right)^{\rho(p) - g} = \left(\frac{2}{3}\right)^{-2}.$$

Finally, by Merten's theorems (see Halberstam and Richert, 1974 p. 35),

$$\sum_{\substack{p < x \\ p \equiv l (\mathrm{mod}\ k)}} \frac{1}{p} = \frac{1}{\varphi(k)} \log \log x + O_k(1), \quad (l, k) = 1.$$

Taking $\delta = \frac{1}{2}$, (7.30) implies Theorem 7.15.

With $m = 5, k = 3$, we have the 1956 conjecture (see Guy, 2004) due to Polish mathematician Waclaw Sierpiński (1882–1969) that the equation

$$\frac{5}{n} = \frac{1}{x} + \frac{1}{y} + \frac{1}{z}$$

has positive integer solutions for every $n > 1$.

Bonnie M. Stewart (1914–1994) proved in 1996 that this conjecture holds for all $n \not\equiv 1 \pmod{278468}$, and verified it numerically for all $n \leq 10^9$.

Neither of the two conjectures just mentioned have yet been proved or disproved. There is a third problem, also unsolved, due to Erdös and Martin Graham. Set $m = n = 1$ in (7.28), with $x_1 < x_2 < \cdots < x_k$. Then the problem is to determine the smallest possible value of X_k for any given positive integer k; writing this minimum value as $m(k)$, some known values are $m(3) = 6$, $m(4) = 12$, $m(12) = 30$, corresponding to

$$1 = \frac{1}{2} + \frac{1}{3} + \frac{1}{6}, 1 = \frac{1}{2} + \frac{1}{4} + \frac{1}{6} + \frac{1}{12},$$

$$1 = \frac{1}{6} + \frac{1}{7} + \frac{1}{8} + \frac{1}{9} + \frac{1}{10} + \frac{1}{14}$$

$$+ \frac{1}{15} + \frac{1}{18} + \frac{1}{20} + \frac{1}{24} + \frac{1}{28} + \frac{1}{3020}$$

$$+ \frac{1}{24} + \frac{1}{28} + \frac{1}{30}.$$

Nobody has been able to provide a general answer to this question however. Erdös asked additionally whether or not there exists a positive constant c such that $m(k) \leq ck$.

Considering the Egyptian fraction problems above, and noting that $5 = 4 + \frac{1}{2} + \frac{1}{2}, 4 \times \frac{1}{2} \times \frac{1}{2} = 1$, it follows that the Sierpiński-type additive and multiplicative equation

$$\begin{cases} \frac{5}{n} = x + y + z, \\ xyz = \frac{1}{A} \end{cases}$$

has a solution for every positive integer n, where x, y, z are positive rational numbers, A a positive integer. However, we have not been able to find any solution in positive rational numbers to

$$\begin{cases} 4 = x + y + z \\ xyz = \frac{1}{A}, \end{cases}$$

nor have we been able to prove that no such solution exists, although we have been able to find solutions relaxing the requirement that every component is positive, for example, $(x, y, z) = (-\frac{1}{6}, -\frac{1}{3}, \frac{9}{2})$. Therefore we cannot resolve the Erdös-Straus-type additive and multiplicative equation

$$\begin{cases} \frac{4}{n} = x + y + z \\ xyz = \frac{1}{A} \end{cases}$$

with A a positive integer for arbitrary positive integers n. We propose this as a weak form of the Erdös-Strauss conjecture; it can also be inferred that the Erdös-Strauss conjecture is more difficult than the Sierpiński conjecture.

We have some related results as follows. For positive integer n, let $A(n)$ denote the smallest positive integer A such that the additive and multiplicative equation

$$\begin{cases} n = x + y + z \\ xyz = \frac{1}{A} \end{cases} \tag{7.31}$$

has a positive rational solution. Since $1 = \frac{1}{3} + \frac{1}{3} + \frac{1}{3}, 2 = 1 + \frac{1}{2} + \frac{1}{2}$, the arithmetic-geometric inequality shows that $A(1) = 27, A(2) = 4$; and since

$$3 = 1 + 1 + 1, \quad 5 = 4 + \frac{1}{2} + \frac{1}{2}, \quad 6 = \frac{9}{2} + \frac{4}{3} + \frac{1}{6},$$

$$9 = \frac{49}{6} + \frac{9}{14} + \frac{4}{21}, \quad 10 = \frac{324}{35} + \frac{25}{126} + \frac{49}{90},$$

we can obtain $A(3) = A(5) = A(6) = A(9) = A(10) = 1$. In fact, when $A = 1, (7.31)$ is equivalent to

$$x^3 + y^3 + z^3 = nxyz,$$

or

$$n = \frac{x}{y} + \frac{y}{z} + \frac{z}{x},$$

where x, y, z are positive integers. According to http://oeis.org/ A072716, the positive integers $n \le 100$ such that this equation has solutions are $n = 3, 5, 6, 9, 10, 13, 14, 17, 18, 19, 21, 26, 29, 30, 38,$ $41, 51, 53, 54, 57, 66, 67, 69, 73, 74, 77, 83, 86, 94$. In general, if n is not a multiple of 4, we can find $A(n)$ using Magma and the theory of elliptic curves. For example $A(7) = 6$, with solution $\left(\frac{4232}{825}, \frac{1875}{1012}, \frac{121}{6900}\right)$. But if n is a multiple of 4, $A(n)$ is not easy to determine.

Looking back at the various issues raised and discussed in the preceding seven chapters, we can see that we have made some important partial progress, but there remains a wealth of fascinating work to be done. We hope that these questions can blossom in beautiful flowers and bear rich fruits. Since the conquest of Fermat's last theorem in 1995, there has been much good news in number theory: first, Catalan's conjecture was proved in 2002, and subsequently proofs were announced for the *abc* conjecture in 2012 and the odd Goldbach conjecture in 2013, although the first of these remains unverified. The year 2013 also saw a major breakthrough towards the twin prime conjecture. But these developments also signify a double-edged sword, and a source of some alarm in the number theory community that there remain among the still outstanding problems fewer and fewer geese capable of laying a golden egg. Hopefully this book may serve as a breath of fresh air, even a drop of fresh blood, capable of giving birth to one or two such goslings.

Bibliography

S. Alaca and K. Williams, *Introductory Algebraic Number Theory*. CUP, Cambridge, 2004.

N. C. Ankeny, Sums of three squares, *Proc. Amer. Math. Soc.*, 8 (1957) 316–319.

L. Bastien, Nombres congruents, *Intermediaire Math.*, 22 (1915) 232–232.

D. A. Beal, *The Beal Prize*, AMS, 2013. http://www.ams.org/profession/prizes-awards/ams-supported/beal-prize.

W. Burnside, On the rational solutions of the equation $x^3 + y^3 + z^3 = 0$, in quadratic fields, *Proc. Lond. Math. Soc.*, 14 (1915), 1–4.

T. Cai, *A Modern Introduction to Classical Number Theory*, World Scientific, Singapore, 2021.

T. Cai, *Perfect numbers and Fibonacci sequences*, World Scientific Press, Singapore, 2022.

T. Cai and D. Chen, A new variant of the Hilbert-Waring problem, *Math. Comput.*, 82(2) (2013) 2333–2341.

T. Cai, D. Chen, Z. Shen, The number of representations for $n = a + b$ with $ab = tc^2$ *or* $\binom{c}{2}$, preprint.

T. Cai, D. Chen and Y. Zhang, A new generalization of Fermat's Last Theorem, *J. Number Theory*, 149(4) (2015), 33–45.

T. Cai, D. Chen, and Y. Zhang, Perfect numbers and Fibonacci primes I, *Int. J. Number Theory*, 11(2015), 159–169.

T. Cai, Additive representations with product equal to a polygonal number, *China Adv. Math.*, 53(3) (2024), 667–671.

T. Cai, L. Wang, and Y. Zhang, Perfect numbers and Fibonacci primes II, *Integers*, 19 (A21), (2019) 1–10.

T. Cai and Y. Zhang, N-tuples of positive integers with the same second elementary symmetric function value and the same product, *J. Number Theory*, 132(9) (2012), 2065–2074.

T. Cai and Y. Zhang, N-tuples of positive integers with the same sum and the same product, *Math. Comput.*, 82(1) (2013), 617–623.

T. Cai and Y. Zhang, A variety of Euler's sum of powers conjecture, *Czechoslovak Math. J.*, 71(146) (2021a), 1099–1113.

T. Cai and Y. Zhang, Euler's conjecture and its variant, *China Adv. Math.*, 50(3) (2021b), 475–479.

T. Cai, Y. Zhang, Congruent numbers on the right trapezoid, 2016. arXiv:1605.06774.

T. Cai, Z. Shen, and P. Yang, On the solution set of additive and multiplicative congruences modulo primes, Filomat, 38(2) (2024), 621–635.

J. Cassels, On a diophantine equation. *Acta Arithmetica*, 6(1) (1960), pp. 47–52.

K. Chakraborty, On the diophantine equation $x + y + z = xyz = 1$. *Annales Univ. Sci. Budapest., Sect. Comp.*, 27 (2007), 145–154.

H. V. Chu, What's special about the perfect number 6? *Amer. Math. Monthly*, 128(1), (2021), 87.

H. V. Chu, On even perfect numbers (II), 2020. arXiv: 2001.08633v1.

H. Cohen, *Number Theory. Vol. I: Tools and Diophantine Equations*, Graduate Texts in Mathematics 239. Springer, NewYork, 2007.

J. P. Cook, *The Mass Formula for Binary Quadratic Forms*, preprint.

H. Darman and A. Granville, On the equations $z^m = F(x, y)$ and $A x^p + B y^p = C z^r$, *Bull. London Math. Soc.*, 27(6) (1995), 513–543.

M. Deléglise, Bounds for the density of abundant numbers, *Exp. Math.*, 7(2) (1998) 137–143,

V. A. Dem'janenko, L. Euler's conjecture (Russian), *Acta Arith.*, 25 (1973/74) 127–135.

L. E. Dickson, *History of the Theory of Numbers* (Volumes I–III), Chelsea, Providence, 2002.

N. Elkies, On $A^4 + B^4 + C^4 = D^4$, *Math. Comput.*, 51 (1988) 825–835.

P. Erdös and J. L. Selfridge, The product of consecutive integers is never a prime, *Illinois J. Math.*, 19 (1975) 292–301.

M. Erickson and A. Vazzana, *Introduction to Number Theory*, Chapman & Hall/CRC, 2006.

R. K. Guy, *Unsolved Problems in Number Theory*, Springer, third edition, 2004, p. D11.

H. Halberstam and H. E. Richert, *Sieve Methods*, Academic Press, New York, 1974.

E. Halberstadt and A. Kraus, Une conjecture de Lebesgue, *J. Lond. Math. Soc.*, 69(02) 2004, 291–302.

G. H. Hardy and E. M. Wright, *An Introduction to the Theory of Numbers*, Oxford University Press, Oxford, 1979.

K. Heegner, Diophantische Analysis und Modulfunktionen, *Math. Z.*, 56(1952), 227–253.

H. A. Helfgott, The ternary Goldbach conjecture is true, 2013. arxiv:1312.7748 [math.NT].

H. A. Helfgott, The ternary Goldbach problem, 2014. arXiv:1404.2224 [math.NT].

L. K. Hua, *Introduction to Number Theory*, Springer, Berlin, 1982.

H. Iwaniec, Almost-primes represented by quadratic polynomials, *Inventiones Mathematicae*, 47(2), (1978), 171–188.

L. W. Jacobi and D. J. Madden, $a^4 + b^4 + c^4 + d^4 = (a + b + c + d)^4$, *The Amer. Math. Soc. Monthly*, 151 (2008), 220–236.

X. Jiang, On the even perfect numbers, *Colloq. Math.*, 154(1), 2018, 131–136.

M. Jones, J. Rouse, Solutions of the cubic Fermat equation in quadratic fields, *Int. J. Number Theory*, 9 (2013), 1579–1591.

K. Kazuya, N. Kurokawa, and T. Saito, *Number Theory (I – II)*, Translations of Mathematical Monographs, *Amer. Math. Soc.*, 2000.

Z. Ke, On the Equation $x^2 = y^n + 1, xy \neq 0$, *J. Sichuan Univ. (Natural Science Edition)*, 1 (1962), 1–6.

Z. Ke and Q. Sun, *Discussion of Indeterminate Equations*, Harbin Institute of Technology Press, 2011.

L. J. Lander and T. R. Parkin, Counterexample to Euler's sum of powers conjecture, *Bull. Amer. Math. Soc.*, 72 (1966), 1079.

H. Li, Waring's Problem for Sixteenth Powers, *Sci. China Ser. A*, 39 (1996) 56–64.

P. Mihailescu, Primary cyclotomic units and a proof of Catalan's conjecture, *J. Reine Angew. Math.* 572 (2004) 167–195.

P. Mihailescu, Around ABC, *Newslett. Eur. Math. Soc.*, 93(3), (2014), 29–35.

L. J. Mordell, On the rational solutions of the indeterminate equations of the third and fourth degrees, *Proc. Cambridge Philo. Soc.*, 21 (1922) 179–192.

L. J. Mordell, The diophantine equation $x^3 + y^3 + z^3 + kxyz = 0$, *Colloque sur la théorie des nombres*, Bruselles, 1955, pp. 67–76.

L. J. Mordell, The congruence $((p - 1)/2)! \equiv \pm 1 (\bmod p)$, *Amer. Math. Monthly*, 68 (1961), 145–146.

L. J. Mordell, *Diophantine Equations*, Academic Press, New York and London, 1969.

A. Murty, 2005. http://Oeis.org/A109909.

M. R. Murty, Introduction to p-adic analytic number theory, *Amer. Math. Soc.*, Providence, 2009.

T. Nagell, *Introduction to Number Theory*, Wiley, New York, 1951.

R. Norrie, in University of Saint Andrews 500th Anniversary Memorial Volume of Scientific Papers, published by the University of Saint Andrews, 1911, p. 89.

S. S. Pillai, On the inequality "$0 < a^x - b^y \leq R$", *J. Indian Math. Soc.*, 19 (1931), 1–11.

H. Qin, Congruent numbers, quadratic forms and K_2, *Mathematische Annalen*, 383(2022), 1647–1686.

P. Ribenboim, *The New Book of Prime Number Records*, New York, Springer, 1995.

T. Ross, A perfect number generalization and some Euclid-Euler type results. *J. Inte. Seq.*, Vol. 27 (2024), Article 29.7.5.

G. Sansone and J. Cassels, *Sur le probleme de M. Werner Mnich.* Acta Arithmetica, 7 (1962), pp. 187–190.

A. Schinzel, Triples of positive integers with the same sum and the same product, *Serdica Math. J.*, 22 (1996), 587–588.

J. Silverman, J. Tate, *Rational Points on Elliptic Curves*, Springer, 2004.

T. Skolem, *Diophantische Gleichungen*, Chelsea, Providence, 1950.

A. Smith, The congruent numbers have positive natural density, 2016. arXiv:1603.08479v2.

Z.-W. Sun, On sums of primes and triangular numbers, *J. Combinat. Number Theory*, 1(1) (2009), 65–76.

R. Taylor, A. Wiles, Ring-theoretic properties of certain Hecke algebras, *Ann. Math.* 141 (1995), 553–572.

Y. Tian, Congruent numbers and Heegner points, *Cambridge J. Math.*, 2(1) (2014), 117–161.

Y. Tian, X. Yuan, and S. Zhang, Genus periods, genus points and congruent number problem. *Asian J. Math.* 21(4), (2017), 721–774.

R. Tijdeman, *On the Equation of Catalan*, Acta Arithmetica, 29, (1976), 197–209.

J. B. Tunnell, A classical Diophantine problem and modular forms of weight 3/2, *Invent. Math.*, 72(1983), 323–334.

M. Ulas: On some Diophantine systems involving symmetric polynomials, *Math. Comput.*, 83 (2014), 1915–1930.

R. C. Vaughan, *The Hardy-Littlewood Method*, Cambridge University Press, 1981.

I. M. Vinogradov, *Fundamentals of the Theory of Numbers*, Publishing House of Technology, 1952.

X. Wang, *Some New Properties of Narayana Sequences*, Master's Thesis, Zhejiang University, 2022.

Y. Wang, The congruent number problem and elliptic curves, *Notices Chinese Math. Soc.*, 90(2), (2004), 1–5.

W. A. Webb, On $4/n = 1/x + 1/y + 1/z$, *Proc. Amer. Math. Soc.* 25 (1970) 578–584.

E. W. Weisstein, *Prime Representation. From Math World*, A Wolfram, 1995. Web Resource. http://mathworld.wolfram.com/PrimeRepresentation.html

A. Wiles, Modular elliptic curves and Fermat's last theorem, *Ann. Math.*, 141(3), (1995), 443–551.

X. Yang, *Proc. Amer. Math. Soc.*, 85(4), (1987), 496–498.

H. Zhong, *Some Problems in Number Theory*, Doctoral Dissertation, Zhejiang University, 2019.

H. Zhong, T. Cai, On the number of representations of $n = a + b$ with ab a multiple of a polygonal number, 2017. arXiv:1709.06334.

H. Zhong and T. Cai, Perfect numbers and Fibonacci primes (III), 2017. arXiv:1709.06337.

Author Index

Subject Index

www.ingramcontent.com/pod-product-compliance
Lightning Source LLC
Chambersburg PA
CBHW050551190326
41458CB00007B/2002